Dietrich Braess
Wolfgang Hackbusch
Ulrich Trottenberg (Eds.)

Advances in Multi-Grid Methods

Notes on Numerical Fluid Mechanics
Volume 11

Series Editors: Ernst Heinrich Hirschel, München
Maurizio Pandolfi, Torino
Arthur Rizzi, Stockholm
Bernard Roux, Marseille

Volume 1 Boundary Algorithms for Multidimensional Inviscid Hyperbolic Flows (Karl Förster, Ed.)

Volume 2 Proceedings of the Third GAMM-Conference on Numerical Methods in Fluid Mechanics (Ernst Heinrich Hirschel, Ed.) (out of print)

Volume 3 Numerical Methods for the Computation of Inviscid Transonic Flows with Shock Waves (Arthur Rizzi / Henri Viviand, Eds.)

Volume 4 Shear Flow in Surface-Oriented Coordinates (Ernst Heinrich Hirschel / Wilhelm Kordulla)

Volume 5 Proceedings of the Fourth GAMM-Conference on Numerical Methods in Fluid Mechanics (Henri Viviand, Ed.) (out of print)

Volume 6 Numerical Methods in Laminar Flame Propagation (Norbert Peters / Jürgen Warnatz, Eds.)

Volume 7 Proceedings of the Fifth GAMM-Conference on Numerical Methods in Fluid Mechanics (Maurizio Pandolfi / Renzo Piva, Eds.)

Volume 8 Vectorization of Computer Programs with Applications to Computational Fluid Dynamics (Wolfgang Gentzsch)

Volume 9 Analysis of Laminar Flow over a Backward Facing Step (Ken Morgan / Jaques Periaux / François Thomasset, Eds.)

Volume 10 Efficient Solutions of Elliptic Systems (Wolfgang Hackbusch, Ed.)

Volume 11 Advances in Multi-Grid Methods (Dietrich Breass / Wolfgang Hackbusch / Ulrich Trottenberg, Eds.)

Volume 12 The Efficient Use of Vector Computer with Emphasis on Computational Fluid Dynamics (Willi Schönauer / Wolfgang Gentzsch, Eds.)

Manuscripts should have well over 100 pages. As they will be reproduced fotomechanically they should be typed with utmost care on special stationary which will be supplied on request. In print, the size will be reduced linearly to approximately 75 %. Figures and diagrams should be lettered accordingly so as to produce letters not smaller than 2 mm in print. The same is valid for handwritten formulae. Manuscripts (in English) or proposals should be sent to the general editor Prof. Dr. E. H. Hirschel, MBB-LKE 122, Postfach 80 11 60, D-8000 München 80.

Dietrich Braess
Wolfgang Hackbusch
Ulrich Trottenberg (Eds.)

Advances in Multi-Grid Methods

Proceedings of the conference held in
Oberwolfach, December 8 to 13, 1984

Springer Fachmedien Wiesbaden GmbH

CIP-Kurztitelaufnahme der Deutschen Bibliothek

Advances in multi-grid methods: proceedings of the conference held in Oberwolfach, December 8–13, 1984 / Dietrich Braess ... (ed.). – Braunschweig; Wiesbaden: Vieweg, 1985.
(Notes on numerical fluid mechanics; Vol. 11)
ISBN 978-3-528-08085-3

NE: Braess, Dietrich [Hrsg.]; GT

Ursprünglich erschienen bei Friedr. Vieweg & Sohn Verlagsgesellschaft mbH, Braunschweig 1985

ISBN 978-3-528-08085-3 ISBN 978-3-663-14245-4 (eBook)
DOI 10.1007/978-3-663-14245-4

PREFACE

During the week of December 8-13, 1984, a conference on *Multi-Grid Methods* was held at the Mathematisches Forschungsinstitut (Mathematical Research Institute) in Oberwolfach. The conference was suggested by the GAMM-Committee "Effiziente numerische Verfahren für partielle Differentialgleichungen". We were pleased to have 42 participants from 12 countries.

These proceedings contain some contributions to the conference. The centre of interest in the more theoretical contributions were exact convergence proofs for multi-grid method. Here, the theoretical foundation for the application of the method to the Stokes equations, the biharmonic equation in its formulation as a mixed finite element problem and other more involved problems were investigated. Moreover, improvements and new attacks for getting quantitative results on convergence rates were reported.

Another series of contributions was concerned with the development of highly efficient and fast algorithms for various partial differential equations. Also in this framework, the Stokes and the biharmonic equations were investigated. Other lectures treated problems from fluid mechanics (as Navier-Stokes and Euler equations), the dam-problem and eigenvalue problems.

The editors would like to thank Professor M. Barner, the director of Mathematisches Forschungsinstitut Oberwolfach for making this conference possible.

D. Braess, Bochum W. Hackbusch, Kiel U. Trottenberg, St. Augustin

CONTENTS

A MIXED VARIABLE FINITE ELEMENT METHOD FOR THE EFFICIENT SOLUTION OF NONLINEAR DIFFUSION AND POTENTIAL FLOW EQUATIONS

O. Axelsson

Department of Mathematics, Toernooiveld, University of Nijmegen, The Netherlands

Abstract

A recently developed method ([1]) for the efficient solution of nonlinear partial differential equations of the form $\frac{\partial}{\partial x}(a_1\frac{\partial u}{\partial x}) + \frac{\partial}{\partial y}(a_2\frac{\partial u}{\partial y}) + f = 0$, where $a_i = a_i(x,y,u,u_x,u_y,\nabla u)$, is further discussed in this paper. The method has applications in many important practical problems.

1. Introduction

We shall consider the numerical solution of

$$-\underline{\nabla}.(D\underline{\nabla} u) = f,\ (x,y) \in \Omega \subset \mathbb{R}^2, \tag{1.1}$$

$u = g_1$ on $\Gamma_D \subset \partial\Omega$, $D\underline{\nabla} u.\hat{\underline{n}} = g_2$ on $\Gamma_N = \partial\Omega \setminus \Gamma_D$, where $D = \text{diag}(a_1,a_2)$, $a_i = a_i(x,y,u,u_x,u_y) > 0, i = 1,2$. Such problems arise in many applications, such as

(i) nonlinear heat conduction, where $a_i = a_i(u)$ and u is the local temperature of body Ω,

(ii) electromagnetic field equations, where $a = a(|\nabla u|^2$, and in addition varies significantly, typically with a factor 10^3, between two areas with different materials (air and iron ore). Here u is the potential.

(iii) potential (i.e. irrotational) flow of an adiabatic gas around an obstacle. Here $\underline{\nabla}.(a\underline{\nabla}\phi) = 0$ and $a = \rho_0[1-\tilde{\gamma}(|\nabla\phi/a_0|^2-M_0^2)]^{1/2\tilde{\gamma}}$, where ρ is the local density of the fluid, $\underline{v} = \underline{\nabla}\phi$ is the velocity, ϕ is the potential function, ρ_0, a_0 and M_0 are the density, speed of sound and mach-numbers, respectively, at inflow of a windtunnel or the far-field values. Further $\tilde{\gamma} = (\gamma-1)/2$, where γ is the ratio of specific heats ($\gamma = 1.4$ in air). We can solve the equation in the normalized variable $\tilde{\phi} = \phi/a_0$ for which $\underline{\nabla}.(\tilde{\rho}\nabla\tilde{\phi}) = 0$, $\tilde{\rho} = [1-\tilde{\gamma}(|\nabla\tilde{\phi}|^2-M_0^2)]^{1/2\tilde{\gamma}}$.

The condition for ellipticity of the problem can be shown to be (see, for instance [7]), $a(z) + 2a'(z)z > 0$, where $a = a(z)$ and $z = |\underline{\nabla}\phi|^2$. If this condition is satisfied for the potential gas flow, the flow is called subsonic. One finds that it is subsonic if

$$\max_{\Omega} |\nabla\tilde{\phi}| \leq (1+\tilde{\gamma}\ M_0^2)\ /\ (1+\tilde{\gamma}).$$

As is wellknown, the upper limit is taken in a point where the local mach-number $M = 1$, which marks the emergence of a supersonic bubble.

In this paper we shall discuss the elliptic (subsonic) case only. Other applications where (1.1) occurs are

(iv) certain free boundary value problems, where in general the boundary between two media (such as ice and water) is unknown. In such a case we can't in general construct a finite element mesh, say with triangles, where the line of discontinuity falls along the sides of triangles. Instead this line passes through the interior of some elements and, as is well-known, this may cause a serious degradation of the accuracy of the finite element (or difference) approximation.

We shall show how the accuracy can be improved by a method of homogenization.

(v) inverse (coefficient identification) problems, i.e. problems where we want to determine the unknown material coefficient function (a) from measurements of the solution.

In this paper we will not further address this problem.

(vi) The new method is also of interest for the solution of (nonlinear) Stokes problems and the related porous medium equation, such as arises in reservoir engineering, for instance.

In the later application, we have to solve the fluid velocity field accurately. One may then first solve an equation for the pressure of the form (see [6]) $\underline{\nabla}(a\underline{\nabla}p) = q$, where $a = k/\mu$, k is the permeability or hydraulic conductivity and μ is the viscosity. The coefficient (a) may vary by several orders of magnitude.

The velocity, which is continuous, is related to the pressure by D'Arcy's law, $\underline{u} = -\frac{k}{\mu}\underline{\nabla}p$. Finally, the concentration of the invading fluid is calculated from an equation

$$\phi\frac{\partial c}{\partial t} = \underline{\nabla}.(D\underline{\nabla}c-\underline{u}c) - q\hat{c}.$$

The velocity can be calculated accurately before the calculation of the concentration at the next time step, by taking proper averages of the values of velocity in the mixed finite element solution to be presented.

In a standard finite element method one first constructs local finite element matrices by numerical integration and then assembles these matrices to a global matrix, of order equal to the number of nodepoints for the unknown function u (excluding the Dirichlet points). This process is costly and must be repeated during each nonlinear iteration step (i.e. during each step to solve the nonlinear algebraic equations one gets after applying the finite element method).

We show that, by use of a certain mixed variable finite element method, we can confine updating of the nonlinearity to an (essentially) diagonal matrix (M). Further, assembly is performed only once and for a discretization (B) of the divergence (and gradient) operators.

This is made possible by use of a certain iterative solution method for the solution of the resulting algebraic system $BM^{-1}(u)B^T\underline{\alpha} = \underline{F}$. The method is globally convergent under quite general conditions of $\underline{a}$.

We also indicate how the discrete approximation in the new method can be more accurate for problems with (almost) discontinuous coefficients, where the discontinuity occurs in the interior of the elements. There are many applications where this technique of homogenization is of great importance. More details on this are found in [1] and [4].

2. The standard finite element method.

Let $V_{N+M} = \mathrm{SPAN}\{\phi_1,\phi_2,\ldots,\phi_{N+M}\}$, where ϕ_i are continuous and piecewise polynomial (and of degree k) Lagrangean basisfunctions, corresponding to a finite element mesh with triangular or quadrilateral elements $\{e_l\}_{l=1}$ where $\Omega = \bigcup_{l=1}^{T}(e_l)$, and T is the number of elements.

Let $u_N = \sum_{j=1}^{N+M} \alpha_j\phi_j$ and $\alpha_i = u_N(P_i) = g_1(P_i)$, for the nodepoints P_i, $i = N+1,\ldots,M$ on Γ_D. The coefficients $\{\alpha_j\}_{j=1}^{N}$ are determinated by the variational formulation (Galerkin method),

$$\int_\Omega D\underline{\nabla}u_N\underline{\nabla}\phi_i d\Omega = \int_\Omega f\phi_i d\Omega + \oint_{\Gamma_N} g_2\phi_i d\Omega$$

or

$$K(u_N,\underline{\nabla}u_N)\underline{\alpha} = \underline{F},$$

where

$$K_{i,j} = \int D\underline{\nabla}\phi_j.\underline{\nabla}\phi_i d\Omega,\quad \underline{\alpha} = (\alpha_1,\alpha_2,\ldots,\alpha_N)^T,$$

$$F^{(i)} = \int_\Omega f\phi_i d\Omega + \oint_{\Gamma_N} g_2\phi_i d\Gamma.$$

In practice, at least where D is variable, we have to calculate first

$$K_{i,j}^{(l)} = \int_{e_l} D\underline{\nabla}\phi_j.\underline{\nabla}\phi_i d\Omega$$

on each element by numerical integration and K (and $\underline{F}$) are calculated by assembly,

$$K_{i,j} = \sum_{l=1}^{T} K_{i,j}^{(l)}.$$

This method suffers from two serious disadvantages.

(i) If D is a function of u (or of $\underline{\nabla}u$), updating of K during the nonlinear iterations means that the numerical integration and assembly process must be repeated.
This is timeconsuming and is frequently the most costly part of the computation.

(ii) If the points of discontinuities are not meshpoints on meshlines, the approximation u_N is less accurate near a point of discontinuity of a, or near points where a has sharp gradients.
In fact, due to the elliptic nature of the problem and due to the large numerical integration errors in these elements, the global accuracy in an averaged l_2-norm, $\Sigma h^{-2}[(u_N-u)(P_i)]^2$, is only $O(h^{1/2})$, where h is the meshsize parameter.

3. The mixed variable formulation

Let $\underline{z} = D\underline{\nabla}u$ and note that in practical problems $\underline{z}$ is continuous even if $\underline{\nabla}u$ is discontinuous. With (1.1) this leads to the coupled system

$$D^{-1}\underline{z} - \underline{\nabla}u = 0$$

$$-\underline{\nabla}.\underline{z} = f,\quad (x,y) \in \Omega$$

$$u = g_1 \text{ on } \Gamma_D \text{ and } \underline{z}.\hat{\underline{n}} = g_2 \text{ on } \Gamma_N.$$

The first boundary condition will still be treated as an essential boundary condition and the second as a natural boundary condition for the variational formulation. This takes then the form

$$\int_\Omega D^{-1}\underline{z}.\underline{\tilde{z}}d\Omega - \int_\Omega \underline{\nabla u}.\underline{\tilde{z}}d\Omega = 0 \quad \forall\ \tilde{z} \in [L^2(\Omega)]^2$$

$$\int_\Omega \underline{z}.\underline{\nabla\tilde{v}}d\Omega = \int_\Omega f\tilde{v}d\Omega + \oint_{\Gamma_N} g_2\tilde{v}d\Gamma \quad \forall\ \tilde{v} \in \mathring{H}^1(\Omega),$$

where $\mathring{H}^1(\Omega) = \{v \in H^1(\Omega);\ v = 0 \text{ on } \Gamma_D\}$ and $H^1(\Omega)$ is the first order Sobolevspace.

We now choose finite element subspaces $V \in \mathring{H}^1(\Omega)$, $W \subset [L_2(\Omega)]^2$ and restrict the variations to these subspaces:

$$\begin{aligned}
&\int_\Omega D^{-1}\underline{z}_h \circ \underline{\tilde{z}}_h d\Omega - \int_\Omega \underline{\nabla u}_h \circ \underline{\tilde{z}}_h d\Omega = 0 \quad \forall\ \tilde{z}_h \in W \\
&\int_\Omega \underline{z}_h \circ \underline{\nabla\tilde{v}}_h d\Omega = \int_\Omega f\tilde{v}_h d\Omega + \oint_{\Gamma_N} g_2\tilde{v}_h d\Gamma \quad \forall\ \tilde{v}_h \in V
\end{aligned} \tag{3.1}$$

Here h is a mesh size parameter, associated with the finite element mesh. Our problem is now:

Find $u_h = \hat{g}_1+v_h$, $v_h \in V$, $\hat{g}_1 \in H^1(\Omega)$ and $\underline{z}_h \in W$ where $\hat{g}_1 = g_1$ on Γ_D and $u_h, \underline{z}_h$ satisfies (3.1).

We shall consider the following two choices of finite element spaces for a triangulation $\bigcup_{l=1}^{T} e_l = \Omega$ of the given region.

(a) $V = \text{SPAN}\{\phi_i\}_{i=1}^N$, where ϕ_i are the usual piecewise linear and continuous basisfunctions,

$W = \text{SPAN}\{\{\psi_4^{(l)}\}_{l=1}^T\}^2$

where $\psi_4^{(l)}$ is piecewise constant, with support on e_l. With $\psi_4^{(l)}$ we associate the center of gravity, $N_4^{(l)}$ of e_l, as nodepoint. (See Figure 3.1a).

(b) $V = \text{SPAN}\{\phi_i\}_{i=1}^N$, where ϕ_i are the usual piecewise quadratic and continuous basisfunctions,

$W = \text{SPAN}\{\{\psi_i^{(l)}\}_{i=4}^6\}$, $l = 1,2,\ldots,T$ and

$$\psi_4^{(l)}(N_j^{(l)}) = \begin{cases} 1, & j = 2,4,3 \\ 0, & j = 5,6 \\ -1, & j = 1 \end{cases}$$

is piecewise linear. $N_j^{(l)}$, $j = 4,5,6$ are the mid-edge points of e_l.

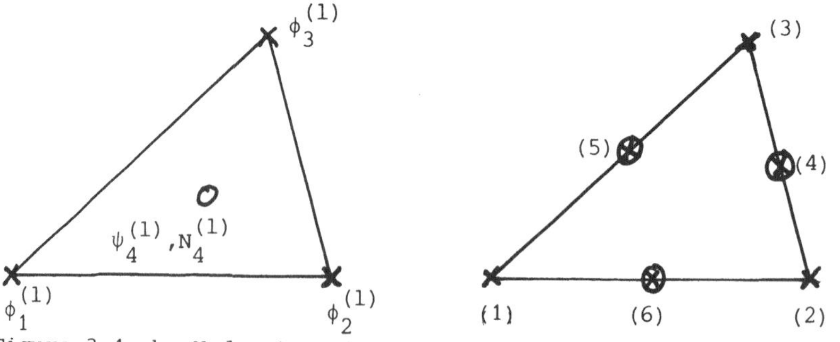

Figure 3.1a,b. Nodepoints in case a,b, respectively; (x) indicates nodepoints for u_h, (O) indicates nodepoints on e_l for $\underline{z}_h$.

Remark 3.1. The global basisfunctions $\{\psi_i\}_{i=1}^{N'}$ have in both cases support on just one element. Hence, at each mid-edge point in case (b), there are two sets of unknowns $\underline{z}_h$, except at the boundary points. It is easy to see that the corresponding linear system takes the form

$$\begin{bmatrix} M^{(1)} & O & -B^{(1)^T} \\ O & M^{(2)} & -B^{(2)^T} \\ B^{(1)} & B^{(2)} & O \end{bmatrix} \begin{bmatrix} \underline{\beta}_1 \\ \underline{\beta}_2 \\ \underline{\alpha} \end{bmatrix} = \begin{bmatrix} O \\ O \\ \underline{F} \end{bmatrix} . \qquad (3.2)$$

Let $e^{(j)} = \text{supp}\ (\psi_j)$, $\Omega^{(i)} = \text{supp}\ (\phi_i)$. Then

$$M^{(k)}_{i,j} = \int_{e^{(j)}} a_k^{-1}\psi_i\psi_j d\Omega, \quad k = 1,2,$$

$$B^{(1)}_{i,j} = \int_{e^{(j)}} \psi_j \frac{\partial}{\partial x}(\phi_i) d\Omega, \ B^{(2)}_{i,j} = \int_{e^{(j)}} \psi_j \frac{\partial}{\partial y}(\phi_i) d\Omega,$$

$$F^{(i)} = \int_{\Omega^{(i)}} f\phi_i d\Omega + \oint_{\bar{\Omega}^{(i)}\cap\Gamma_N} g_2\phi_i d\Gamma .$$

Remark 3.2. $B^{(1)}$ and $B^{(2)}$ are independent of the material property a_i, and depend only on the finite element mesh. Hence they can be calculated once and for all before we start the iterations.
The matrix $B = [B^{(1)},B^{(2)}]$ is in fact a discretization of the divergence operator.

Equation (3.2) can be written,

$$\begin{bmatrix} M & -B^T \\ B & 0 \end{bmatrix} \begin{bmatrix} \underline{\beta} \\ \underline{\alpha} \end{bmatrix} = \begin{bmatrix} \underline{0} \\ \underline{F} \end{bmatrix} .$$

It is easy to see that M is nonsingular. By the elimination of $\underline{\beta}$ we get

$$BM^{-1}B^T\underline{\alpha} = F$$

or

$$[B^{(1)}M^{(1)^{-1}}B^{(1)^T} + B^{(2)}M^{(2)^{-1}}B^{(2)^T}]\underline{\alpha} = \underline{F}.$$

Remark 3.3. The variable $\underline{z}$ is only an intermediate variable. However we may calculate the values of $\underline{z}$ for the nodepoints of each element, associated with $\underline{z}$, that is, $\underline{\beta} = M^{-1}B^T\underline{\alpha}$. Then we may take averages of these values of two (or more) elements, to form more accurate approximations of $z_i = a_i \frac{\partial u}{\partial x_i}$ at some points. We shall not discuss this further in this report.

Numerical integration

For the evaluation of $M^{(k)}_{i,j}$, k = 1,2, we have in general to use numerical integration, when D' is not constant. Hence, the entries of M will be approximate. The entries of $B^{(k)}$ can however be calculated exactly, for instance by the use of the quadrature formulas below.

Theorem 3.1. If we use the following quadrature formulas, in case (a); $\int_{e_1} f d\Omega \cong \text{area}\ (e_1)\ f(N_4^{(1)})$,

where $N_4^{(1)}$ is the center of gravity of $e^{(1)}$, and in

case (b), $\int_{e_1} f d\Omega \cong \frac{1}{3}\ \text{area}\ (e_1) \sum_{j=4}^{6} f(N_j^{(1)})$, where $N_j^{(1)}$, j = 4,5,6 are the midedge points, for the numerical evaluation of $M_{i,j}^{(k)}$, then the corresponding matrices $M^{(k)}$, k = 1,2 becomes diagonal.

Proof. case (a): In this case the statement follows directly, because $\mathrm{supp}(\psi_i) \cap \mathrm{supp}(\psi_j) = \phi$ (the empty set), if $i \neq j$.
case (b): We have only to prove that for $\psi_i^{(1)}$, $\psi_j^{(1)}$, belonging to the same element e_1,

$$M_{i,j}^{(k)} = \int_{e_1} \frac{1}{a_k} \psi_i \psi_j d\Omega = 0,\ i \neq j. \text{ We get } M_{i,j}^{(k)} \cong \frac{1}{3} \operatorname{area}(e_1) a_k^{-1}(N_j^{(1)}) \delta_{i,j},$$

where $\delta_{i,j}$ is the Kroneckersymbol, because $\psi_i \psi_j = 0$, at $N_j^{(1)}$, if $i \neq j$. ∎

In the following we use $M_{i,j}$ to denote the entries we get after numerical integration.

Remark 3.4. Note that updating of the material coefficients a_k affects only M, and is simple when M is diagonal.

Remark 3.5. The same trick works also for quadrilaterals, where the reference element is rectangular. We then use Lobatto quadrature points as nodepoints for ϕ_i and Gauss-points for ψ_i. The numerical integration utilizes the Gauss-points.

Remark 3.6. For elements where a_k is discontinuous, it is advisable to perform the integration more accurately. Then in general, $M^{(k)}$ is not diagonal for rows, corresponding to nodepoints belonging to such elements.

The resulting entries of ${M^{(k)}}^{-1}$ are now weighted harmonic means of a_k. This results in much more accurate approximations of u. For a further discussion, see [4] and [1].

4. Discretization errors

We first prove (compare also [1]) that the mixed variable formulation presented in Section 3, is identical with the classical finite element method when we use the numerical integration methods (cases a,b) as described. Note, however, that in order to achieve a better accuracy, we need a more accurate numerical integration method in elements where a_k has sharp gradients. (See Remark 3.6.)

We then prove that for monotone and Lipschitzbounded operators we have a quasioptimal rate of convergence of the discretization errors.

Theorem 4.1. The finite element mixed variable formulation is identical with the classical finite element formulation, if one uses for case (a), the one-point quadrature method and, for case (b), the three-point quadrature method.

Proof. Consider case (b). (Case (a) follows even easier.)
In the classical formulation we have

$$K_{i,j} = \int_\Omega a \underline{\nabla} \phi_j \nabla \phi_i d\Omega = K_{i,j}^{(1)} + K_{i,j}^{(2)},$$

where

$$K_{i,j}^{(1)} = \int_\Omega a \frac{\partial}{\partial x} \phi_i \frac{\partial}{\partial x} \phi_j d\Omega,\ K_{i,j}^{(2)} = \int_\Omega a \frac{\partial}{\partial y} \phi_i \frac{\partial}{\partial y} \phi_j d\Omega.$$

With the three point quadrature rule we get

$$K^{(1)}_{i,j} = \sum_{l=1}^{T} \frac{1}{3} \text{area}(e_l) \sum_{m=4}^{6} a(N^{(l)}_m)\frac{\partial}{\partial x}\phi_i(N^{(l)}_m)\frac{\partial}{\partial x}\phi_j(N^{(l)}_m). \quad (4.1)$$

Note that when a is constant, the numerical integration is exact, because $\frac{\partial \phi_i}{\partial x}$ is a piecewise linear polynomial and the quadrature is exact for full quadratic polynomials in x and y.
A similar expression is valid for $K^{(2)}_{i,j}$.
In the mixed variable formulation we have

$$\tilde{K} = \tilde{K}^{(1)} + \tilde{K}^{(2)}, \text{ where}$$
$$\tilde{K}^{(i)} = B^{(i)} M^{(i)^{-1}} B^{(i)^T}, \; i = 1,2.$$

Hence, because $M^{(i)}$ is diagonal, we have

$$\tilde{K}^{(1)}_{i,j} = \sum_{k=1}^{N'} B^{(1)}_{i,k} M^{(1)^{-1}}_{k,k} B^{(1)^T}_{j,k} =$$
$$= \sum_{k=1}^{N'} \{\int_\Omega \psi_k \frac{\partial}{\partial x}\phi_i d\Omega [\int_\Omega a^{-1}\psi_k^2 d\Omega]^{-1} \int_\Omega \psi_k \frac{\partial}{\partial x}\phi_j d\Omega\}. \quad (4.2)$$

Here N' is the dimension of the matrix M.
Note that $\text{supp}(\psi_k)$ is one single element, $e^{(k)}$. We now use the three point quadrature rule for the evaluation of the integrals in (4.2). Note then that for $\int_\Omega \psi_k \frac{\partial}{\partial x}\phi_i d\Omega = \int_{e^{(k)}} \phi_k \frac{\partial}{\partial x}\phi_i d\Omega$, this is exact, because ψ_k and $\frac{\partial}{\partial x}\phi_i$ are linear in x and y. If a is constant, $\int_{(k)} a^{-1}\psi_k^2 d\Omega$ will also be evaluated without any numerical integration error.
We get

$$\tilde{K}^{(1)}_{i,j} = \sum_{k=1}^{N'} \frac{1}{3}\text{area}(e^{(k)})\frac{\partial}{\partial x}\phi_i(N^{(k)})[\frac{1}{3}\text{area}(e^{(k)})a(N^{(k)})^{-1}]^{-1}$$
$$\frac{1}{3}\text{area}(e^{(k)})\frac{\partial}{\partial x}\phi_j(N^{(k)}),$$

where $\{N^{(k)}\}_{k=1}^{N'}$ is the set of quadrature points. Note then that these points appear twice (except the boundary quadrature points).
It follows that

$$\tilde{K}^{(1)}_{i,j} = \sum_{k=1}^{N'} \frac{1}{3}\text{area}(e^{(k)})a(N^{(k)})\frac{\partial}{\partial x}\phi_i(N^{(k)})\frac{\partial}{\partial x}\phi_j(N^{(k)}).$$

A similar expression is valid for $\tilde{K}^{(2)}_{i,j}$.
Comparing this with (4.1) we find,

$$\tilde{K}^{(k)}_{i,j} = K^{(k)}_{i,j}, \; k = 1,2.$$

Since the righthand side is the same in both cases, we have the same discretization. ∎

Consider now the discretization error estimate for the nonlinear problem,

$$F(u) = -\underline{\nabla}.(a\underline{\nabla}u) = f, \quad (4.3)$$

with (for notational simplicity only) homogeneous boundary conditions. Here $a = a(x,y,u,u_x,u_y)$.
Note first that boundedness (and hence stability) of the solution follows from

$$\int_\Omega a|\underline{\nabla}u|^2 d\Omega = \int fud\Omega,$$

i.e. because $a \geq a_0 > 0$, $\int_\Omega |\underline{\nabla}u|^2 d\Omega \leq \frac{1}{a_0} ||f||_{V'}$, or $||u||_V \leq C||f||_{V'}$, for some constant C where $V = \overset{\circ}{H}{}^1(\Omega)$ is the Sobolev space and V' is its dual space. The same bound is valid for the discrete solution u_h, which satisfies

$$(F(u_h),v_h) = \int_\Omega a\underline{\nabla}u_h\underline{\nabla}v_n d\Omega = \int_\Omega fv_h d\Omega \ \forall v_h \in V_h \subset V \quad , \qquad (4.4)$$

where V_h is the finite element space. Here $a = a(x,y,u_h,(u_h)_x,(u_h)_y)$. Let $(.,.)$ be the inner product in $L_2(\Omega)$.

We now assume that the problem (4.3) is monotone, i.e.

$$(F(u)-F(v),u-v) \geq \alpha||u-v||_V^2, \ \alpha > 0 \ \forall u,v \in V. \qquad (4.5)$$

If $a = a(u)$, it is easy to see that this is the case if a is Lipschitzcontinuous, $|a(u)-a(v)| \leq L|u-v|$, where L is small enough.

If $a = a(|\underline{\nabla}u|^2)$, it is wellknown that the problem is monotone if $a(z) + 2a'(z)z \geq a_1 > 0$, where $z = |\underline{\nabla}u|^2$, see, for instance [7].

We also assume that the operator L is Lipschitzcontinuous, i.e.

$$||F(u)-F(v)||_{V'} \leq \Gamma||u-v||_V \quad \forall u,v \in V \qquad (4.6)$$

for some constant Γ. This follows if for instance a is Lipschitzcontinuous in its arguments $u,\underline{\nabla}u$.

The discretization error estimate now follows easily. Note at first that by (4.3), (4.4), $(F(u)-F(u_h),v_h) = 0 \quad \forall v_h \in V_h$.
Then using monotonicity (4.5) and the Lipschitz-bound (4.6), we get

$$\alpha||u-u_h||_V^2 \leq (F(u)-F(u_h),u-u_h) = (F(u)-F(u_h),u-v_h)$$

$$\leq \Gamma||u-u_h||_V||u-v_h||_V \quad \forall v_h \in V_h,$$

i.e.

$$||u-u_h||_V \leq \frac{1}{\alpha}\Gamma \inf_{v_h \in V_n} ||u-v_h||_V.$$

Hence, the discretization error is quasioptimal, i.e. never worse than the constant Γ/α times the best approximation of u by elements in V_h.

It remains to estimate the effect of quadrature errors. For a discussion, see [5] and [2].
It turns out that in case (a), $||u-u_h||_V = O(h^r)$, if $u \in H^{r-1}(\Omega) \cap \overset{\circ}{H}{}^1(\Omega)$, $r = 2,3$, respectively.

5. Iterative method

We shall now present a globally convergent class of iterative methods for the solution of the variational equation

$$(F(u_h),v_h) = (f,v_h) \quad \forall v_h \in V_h. \qquad (5.1)$$

To this end we shall consider a "preconditioned" fixed-point iteration (modified Picard iteration) method. Hence, let

$$F(u) = A(u)u = f$$

and

$$F(u_h) = A(u_h)u_h \quad ,$$

where $A(u) = -\frac{\partial}{\partial x} a_1 \frac{\partial}{\partial x} - \frac{\partial}{\partial y} a_2 \frac{\partial}{\partial y}$, where $a_i(x,y,u,u_x,u_y)$, and consider the iterative method

$$A(u^{(0)})(u^{(m+1)}-u^{(m)}) = \tau(f-A(u^{(m)})u^{(m)}), \; m = 0,1,\ldots \tag{5.2}$$

where $u^{(0)}$ is an arbitrary function in V satisfying the essential (i.e. Dirichlet) boundary conditions on $\partial\Omega$. Here $\tau > 0$ is a parameter, which we shall choose small enough to guarantee global convergence, i.e. convergence for any initial function $u^{(0)} \in V_h$ to the solution of (5.1). Actually, the method also proves the existence of a solution to $F(u) = f$ and to (5.1).
Let more generally

$$A_0(u^{(m+1)}-u^{(m)}) = \tau(f-A(u^{(m)})u^{(m)}),$$

where $A_0 : V \to V'$ is any invertible linear operator $(A_0^{-1} : V' \to V)$.
Then

$$u^{(m+1)} = u^{(m)} - \tau[A_0^{-1}A(u^{(m)})u^{(m)}-A_0^{-1}f], \; m = 0,1,\ldots$$

or

$$u^{(m+1)} = T(u^{(m)}), \text{ where} \tag{5.3}$$

$T(u) = Iu - \tau[A_0^{-1}A(u)u-A_0^{-1}f]$ and I is the identity operator. As well known, (5.3) converges for any $u^{(0)}$ if the mapping $T : V \to V$ is contractive, i.e. if $(T(u)-T(v),T(u)-T(v)) \leq \sigma(u-v,u-v)$, $\forall u,v \in V$, where $0 \leq \sigma < 1$ and (,) is the inner product in $L_2(\Omega)$. We have

$$\begin{aligned}(T(u)-T(u),T(u)-T(v)) &= (u-v,u-v) - 2\tau(A_0^{-1}(F(u)-F(v)),u-v) \\ &+ \tau^2||A_0^{-1}(F(u)-F(u))||^2 .\end{aligned} \tag{5.4}$$

We now assume that A_0 is such that $A_0^{-1}F$ is a monotone operator, i.e., that

$$(A_0^{-1}(F(u)-F(v)),u-v) \geq \alpha||u-v||^2, \; \alpha > 0 \; \forall u,v \in V.$$

By (4.5), this is valid if $A_0 = I$, the identity operator, but we get in general larger values of α if we choose $A_0 = A(u^0)$, for instance. Furthermore $||A_0^{-1}(F(u)-F(v)|| \leq M||u-v|| \; \forall u,v \in V$, because by (4.6), F is Lipschitz-bounded. The constant M can be expected to be smaller if we choose A_0 properly.
By (5.4), we get now

$$(T(u)-T(v),T(u)-T(v)) \leq [1-2\tau\alpha+\tau^2M^2]||u-v||^2 .$$

Here $1-2\tau\alpha + \tau^2M^2 = 1-\tau(2\alpha-\tau M^2) < 1$ if $0 < \tau < 2\alpha/M^2$. Hence, for any such value of τ we get global convergence.
By proper choices of A_0 we can expect α and M to be close to 1.
The choice $A_0 = A(u^0)$ gives a modified fixed-point iteration, the choice $A_0 = A(u^{(m)})$ gives a dynamically modified fixed-point iteration (which case the above theory, however does not cover). Numerical tests in [1] indicate very fast convergence.
The variational formulation of the modified fixed-point iteration method is

$$\int_\Omega a^{(0)}\underline{\nabla}(u^{(m+1)}-u^{(m)}).\underline{\nabla}v_h d\Omega = \tau[\int_\Omega fv_h d\Omega - \int_\Omega a^{(m)}\underline{\nabla}u^{(m)}.\underline{\nabla}v_h d\Omega] \; \forall v_h \in V_h,$$

$m = 0,1,\ldots$, where $a^{(m)} = a(x,y,u^{(m)},\underline{\nabla}u^{(m)})$ (assuming for notational simplicity that $a_1 = a_2 = a$).

When we use the mixed variable formulation, we get

$$BM^{(0)^{-1}}B^T(\alpha^{(m+1)}-\alpha^{(m)}) = \tau(F-BM^{(m)^{-1}}B^T\alpha^{(m)}), \; m = 0,1,\ldots, \tag{5.5}$$

where $M^{(m)} = M(u_h^{(m)})$. Note that the matrix $BM^{(m)^{-1}}B^T$ does not have to be formed explicitly, because we need only to perform matrix vector multiplications with it.

Spectral equivalence.

As is shown in [1], to make the method still more computationally efficient, we can use a higher order (i.e. the piecewise quadratic linear combination) finite element approximation for the residuals (the right-hand side of (5.5)), but let A_0 be based on the lowest order (i.e. piecewise linear-constant combination). We then get

$$\tilde{B}\tilde{M}^{(0)^{-1}}\tilde{B}^T(\alpha^{(m+1)}-\alpha^{(m)}) = \tau(F-BM^{(m)^{-1}}B^T\alpha^{(m)}), \quad m = 0,1,\ldots, \tag{5.6}$$

where $\tilde{B}$, $\tilde{M}$ are calculated from the piecewise linear-constant basisfunctions, corresponding to the same nodes as for the piecewise quadratic-linear basisfunctions. Such methods are called spectrally equivalent preconditionin See further [1] and [2].

At every step of (5.6) we have to solve a linear system with matrix $A^{(0)} = \tilde{B}\tilde{M}^{(0)^{-1}}\tilde{B}^T$. This matrix is then calculated explicitly. It is hence somewhat costly to update the "preconditioning", i.e. to recalculate $\tilde{B}\tilde{M}^{(0)^{-1}}\tilde{B}^T$ with a newer vector $u^{(0)}$. For the solution of the linear systems with $A^{(0)}$, we may for instance use incomplete factorization methods accelerated with a conjugate gradient method, or we may use a multigrid method.

Generalized inverse preconditioning.

Consider now an alternative method, based on an approximation of the inverse $\tilde{B}\tilde{M}^{(0)^{-1}}\tilde{B}^T$. It is proven in [3] that

$$(\varepsilon I+\tilde{B}\tilde{M}^{-1}\tilde{B}^T)^{-1} = (\tilde{B}\tilde{B}^T)^{-1}\tilde{B}[\tilde{M}-\tilde{M}(\tilde{M} + \frac{1}{\varepsilon}\tilde{B}^T\tilde{B})^{-1}\tilde{M}]\tilde{B}^T(\tilde{B}\tilde{B}^T)^{-1}, \tag{5.7}$$

where $\varepsilon > 0$. Hence we use the approximation

$$(\tilde{B}\tilde{M}^{-1}\tilde{B}^T)^{-1} = (\tilde{B}\tilde{B}^T)^{-1}\tilde{B}\tilde{M}\tilde{B}^T(\tilde{B}\tilde{B}^T)^{-1}.$$

This approximation is especially accurate when $\tilde{M}$ is close to a constant times the identity matrix. The corresponding iterative method becomes

$$\alpha^{(m+1)} = \alpha^{(m)}+\tau(\tilde{B}\tilde{B}^T)^{-1}\tilde{B}\tilde{M}^{(0)}\tilde{B}^T(\tilde{B}\tilde{B}^T)^{-1}(F-BM^{(m)^{-1}}B^T\alpha^{(m)}), \tag{5.8}$$

$$m = 0,1,\ldots$$

A multiplication by $(\tilde{B}\tilde{B}^T)^{-1}$ is performed by solving a linear system with $\tilde{B}\tilde{B}^T$. Note then, that by the result in Section 4, $\tilde{B}\tilde{B}^T$ is identical to the finite element matrix for a <u>constant coefficient</u> Poisson problem. Hence it is close (or even identical for certain meshes) to the central difference matrix. There exist many extremely fast Poisson solvers for such problems, such as FFT (fast Fourier transforms), cyclic reduction ordering methods and, for more general meshes, the multigrid methods. Also blockwise incomplete factorization methods have been shown to yield very fast methods.

Since every iteration in (5.8) needs only two such Poisson solvers plus some matrix-vector multiplications with sparse matrices, the cost per iteration is small. In addition, the number of iterations are few. Note that we don't have to assemble $M^{(m)}$ (because it is diagonal) and updating of it is hence inexpensive. Since dynamic updating of the preconditioning costs little in (5.8) we may let $\tilde{M}^{(0)} = \tilde{M}^{(m)}$, with negligible extra cost per iteration. Hence this is to be recommended.

The choice $0 < \tau < 1$ of τ corresponds to damped iterations, to get global convergence. Since we can use dynamic updating, we can expect that τ can be chosen quite close to 1, or even larger than 1 (overrelaxation) without getting into divergence. We may even in some problems choose

$\tau = \tau_m$ dynamically, in order to accelerate the rate of convergence.

Nonlinear boundary conditions and nonstationary problems.

Note that $D = D(u)$, so the Neuman-type boundary condition in (1.1) is nonlinear. If in addition $g_2 = g_2(u)$, we just let the residual (on variational form) be defined by

$$(F(u_h), v_h) = \int_\Omega D(u_h)\underline{\nabla}u_h . \underline{\nabla}v_h d\Omega - \oint_{\Gamma_N} g_2(u) v_h d\Gamma \; \forall v_h \in V_h,$$

but the preconditioning can be defined as before.

For a nonstationary problem,

$$c\frac{\partial u}{\partial t} = \underline{\nabla}.(D\underline{\nabla}u) - f \;,$$

where $c = c(u) > 0$, and where we use an implicit timestepping method, we have to solve an algebraic system of the form, like

$$(\frac{1}{\theta k} C + BM^{-1}B^T)\alpha(t+h) = \ldots$$

for some given righthand side at every timestep. Here C is similar to a massmatrix, k is the timestep and $0 < \theta \leq \frac{1}{2}$ is a parameter.

For this, we calculate a preconditioning

$$A^{(0)} = \frac{1}{\theta k} \tilde{C}^{(0)} + \widetilde{\widetilde{BM}}^{(0)^{-1}} \tilde{B}^T, \quad \text{with } u = u^{(0)} .$$

Frequently, this can be kept for several timesteps.

Again, updating of the residual costs little.

Methods of the type (5.6) and (5.8) are also of interest for the solution of Stokes and Navier-Stokes problems and for Navier's equations for almost incompressible materials, where one gets a matrix of the form $\varepsilon M + B^T B$.

Research about these latter applications will be reported elsewhere.

REFERENCES

[1] O. AXELSSON and I. GUSTAFSSON, An efficient finite element method for nonlinear diffusion problems, submitted.

[2] O. AXELSSON and V.A. BARKER, Finite Element Solution of Boundary Value Problems. Theory and Computation. Academic Press, Orlando, 1984.

[3] O. AXELSSON, Numerical Algorithms for indefinite problems, in Elliptic Problem Solvers II, (G. Birkhoff and A. Schoenstadt, eds.), Academic Press, 1984.

[4] I. BABUŠKA and J. OSBORN, Generalized finite element methods: Their performance and their relation to mixed methods, SIAM J. Numer. Anal. 20 (1983), 510-536.

[5] P. CIARLET, The Finite Element Method for Elliptic Problems. North-Holland Publ., Amsterdam, 1978.

[6] R.E. EWING (editor), The Mathematics of Reservoir Simulation, SIAM, Philadelphia, 1984.

[7] M.M. VAINBERG, Variational method and method of monotone operators in the theory of nonlinear equations, John Wiley, New York, 1973.

TWO MULTI-LEVEL ALGORITHMS FOR THE DAM PROBLEM

C. Bollrath

Mathematisches Institut, Ruhr-Universität
D-4630 Bochum 1, Federal Republic of Germany

SUMMARY

We describe two multi-level algorithms for the numerical solution of stationary porous flow free boundary problems. The first one calculates two convergent sequences of super - solutions and subsolutions. It combines projected relaxation steps as proposed by Alt in [2] with corrections of the pressure in the saturated region. This conservative correction preserves the monotonicity. In the second (heuristic) algorithm, we apply the FAS technique[5] to the dam problem and use an approximation of the full problem on the coarse grid. Severalnumerical examples are presented. For large problems, the multi-level algorithms are significantly faster than previous algorithms in which only one grid is used.

1. INTRODUCTION

The study of the stationary flow of an incompressible fluid through a porous medium leads to an elliptic free boundary problem, called the dam problem. The first rigorous mathematical treatment of this problem was carried out by Baiocchi [3]. In the special case of a rectangular domain, he transformed the dam problem into a variational inequality of obstacle type. Recently, Brandt, Cryer [6] and Mandel[8] presented very efficient multigrid solvers for this problem.

However, the application of Baiocchi's approach requires severe restriction on the geometry of the domain. Later, Alt[1] , and Brezis, Kinderlehrer and Stampacchia[7] introduced a new formulation and treated general domains. Here, the pressure u and a second unknown γ are described as the solution of a more complicated variational inequality. The function γ turns out to be, in most of the practical cases, the characteristic function of the set $\{u>0\}$. Nevertheless, in some situations the region $\{0<\gamma<1\}$ is not empty and may be interpreted as an unsaturated zone. The first numerical approach in this framework is also due to Alt[2]. He established, under suitable assumptions, that at least a subsequence of the finite element solutions converges to a solution of the continuous problem. The discrete problem is solved by applying a Jacobi type relaxation method with projection. In the case of a fine discretization this method converges very slowly. Marini and Pietra[9] improved the relaxation methods. The most efficient among their algorithms is the heuristic "Incorporated SORP" (systematic overrelaxation with projection), which uses the relaxation parameter $\omega = 1$ in the unsaturated region $\{\gamma < 1\}$ and ω close to 2 in the wet region of the porous medium.

In the present paper we describe two ways of applying the multigrid idea to the general dam problem. Several numerical tests (see § 5) show that our multi-level algorithms are more efficient than comparable relaxation methods.

2. STATEMENT OF THE PROBLEM AND DISCRETIZATION

We consider the seepage of water through a dam. Let the 2-dimensional section Ω of the dam be a polygonal domain. We denote by Γ^+ the part of the boundary $\partial\Omega$ in contact with water reservoirs and by Γ^o the part in contact with the air. The third part $\partial\Omega\setminus(\Gamma^+ \cup \Gamma^o)$ is assumed to be impervious. The sets Γ^+, Γ^o are measurable and disjoint, the measure of $\Gamma^o \cup \Gamma^+$ is positive.

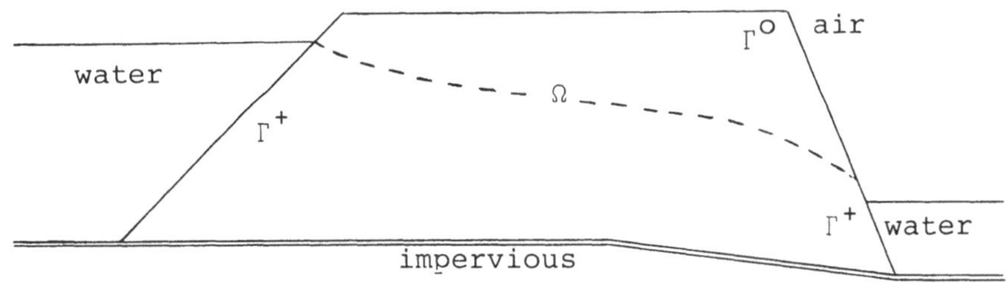

Fig. 1

The permeability of the soil is given by a matrix function

$$a \in L^\infty(\Omega), \quad a \text{ strictly elliptic.}$$

For the special discretization described below, we assume that the medium is isotropic. The Dirichlet boundary values of the pressure are specified by the function $u^o \in C^o(\overline{\Gamma^o \cup \Gamma^+})$, which is positive on Γ^+ and zero on Γ^o. The vertical unit vector $\underline{e}$ gives the direction of gravity.

According to [1] and [7], the flow of water through Ω is described by a variational inequality:

Find $u \in M(u^o) := \{v \in H^1(\Omega),\ v = u^o \text{ on } \Gamma^+,\ v \leqq u^o \text{ on } \Gamma^o\}$ and $\gamma \in L^\infty(\Omega)$ with

$$u \geqq 0, \quad 0 \leqq \gamma \leqq 1, \quad \gamma = 1 \text{ on } \{u > 0\},$$

such that (2.1)

$$\int_\Omega \nabla(v-u)\, a\, (\nabla u + \gamma \underline{e}) \geqq 0$$

holds for every $v \in M(u^o)$.

We assume that there is a uniform triangulation Ω_o of Ω, which is of the following type:

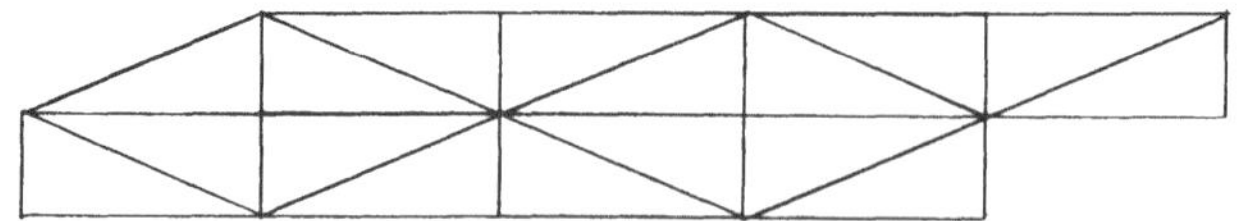

Fig. 2 Section of the triangulation.

The triangulation has to be constructed such that every $p\in\Gamma$ in which the type of the boundary condition changes is a nodal point of Ω_o. Multi-level solution processes typically employ a sequence of increasingly finer grids $\Omega_1,\ldots,\Omega_{k_{max}}$, with mesh sizes $h_k=h_{k-1}/2$. The sets of nodal points are denoted by $P_k, k\in\{0,\ldots,k_{max}\}$.

We set

$$P_k^+ := P_k \cap\overline{\Gamma}^+, P_k^o := (P_k\cap\overline{\Gamma}^o)\setminus P_k^+,\ D_k := P_k\setminus(P_k^o \cup P_k^+).$$

In order to define the f.e. space for γ, we introduce the up-wind triangle associated with the nodal point $p_i\in P_k$:

A triangle $T\in\Omega_k$ is called an "*up-wind triangle*" of p_i, if

i) p_i is a vertex of T and

ii) $T\setminus\{p_i\}$ intersects the oriented half line with end point p_i and direction $\underline{e}$.

Let χ_k^i be the characteristic function of the union of the up-wind triangles associated with the nodal point p_i and let L_k be the linear space spanned by χ_k^i, $p_i\in P_k$, i.e.

$$L_k := \{\gamma_k\in L^\infty(\Omega), \gamma_k(x) = \sum_{i=1}^{n_k} \gamma_k^i\ \chi_k^i(x)\},\ n_k := \mathrm{card} P_k.$$

For a nodal point $p_i\in P_k$ let φ_k^i denote the continuous, on Ω_k piecewise linear function with $\varphi_k^i(p_i)=1$ and $\varphi_k^i(p_j)=0$ for points $p_j\neq p_i$.

The pressure u is approximated in the finite dimensional space

$$S_k := \{u_k\in C^o(\Omega),\ u_k(x) = \sum_{i=1}^{n_k} u_k^i\ \varphi_k^i(x)\}.$$

The corresponding nodal vectors $(\gamma_k^i)_{i=1}^{n_k}$ and $(u_k^i)_{i=1}^{n_k}$ are denoted by $\underline{\gamma}_k$ and $\underline{u}_k$.

We approximate the permeability by a matrix function

$a_h\in L^\infty(\Omega)$, a_h strictly elliptic and constant on every triangle of Ω_o.

The matrices A_k and E_k are defined as follows

$$A_k := (a_{ij}^k)_{i,j=1}^{n_k}, \quad a_{ij}^k := \int_\Omega \nabla\varphi_k^i \, a_h \nabla \varphi_k^j ,$$

$$E_k =: (e_{ij}^k)_{i,j=1}^{n_k}, \quad e_{ij}^k := \int_\Omega \nabla\varphi_k^i \, a_h \chi_k^j \, \underline{e} .$$

These definitions lead to the following discretization of (2.1):

Find $u_k \in M_k(u^o) := \{v_k \in S_k, v_k^i = u^o(p_i)$ for $p_i \in P_k^+, v_k^i \leqq u^o(p_i)$ for $p_i \in P_k^o\}$ and $\gamma_k \in L_k$ with

$$u_k^i \geqq 0, \quad 0 \leqq \gamma_k^i \leqq 1, \quad (1-\gamma_k^i) u_k^i = 0 \quad \text{for } p_i \in P^k, \tag{2.2a}$$

such that

$$(A_k \underline{u}_k + E_h \underline{\gamma}_k)^i \begin{cases} = 0 & \text{for } p_i \in D_k, \\ \leqq 0 & \text{for } p_i \in P_k^o. \end{cases} \tag{2.2b}$$

3. A TWO-LEVEL ALGORITHM OF MONOTONIC TYPE

In order to solve the discrete problem (2.2), Alt introduced in [2] a Jacobi type relaxation operator with projection:

$$R_k : S_k \times L_k \longrightarrow S_k \times L_k, \quad R_k := (R_k^u, R_k^\gamma),$$

$$(R_k^u(u_k,\gamma_k))^i := \begin{cases} \max\{0, (C^i(u_k,\gamma_k) - e_{ii}^k)/a_{ii}^k\} & \text{for } p_i \in D_k, \\ u_o(p_i) & \text{otherwise}, \end{cases}$$

$$(R_k^\gamma(u_k,\gamma_k))^i := \begin{cases} \min\{1, C^i(u_k\gamma_k)/e_{ii}^k\} & \text{for } p_i \in P_k \setminus P_k^+ \text{ with } e_{ii}^k \neq 0, \\ 1 & \text{otherwise} \end{cases}$$

$$C^i(u_k,\gamma_k) := (\underline{b}_k - A_k \underline{u}_k - E_k \underline{\gamma}_k)^i + a_{ii}^k u_k^i + e_{ii}^k \gamma_k^i . \tag{3.1}$$

Note, that we allow in (3.1) (for later use) a non-trivial right hand side $\underline{b}_k \in R^{nk}$. Since the medium is isotropic (i.e. $a_h = d\,Id, d:\Omega \to \mathbb{R}$), we obtain the relations

$$\begin{aligned} a_{ii}^k > 0, \quad a_{ij}^k \leqq 0 \text{ for } i \neq j \\ e_{ii}^k \geqq 0, \quad e_{ij}^k \leqq 0 \text{ for } i \neq j \end{aligned} \tag{3.2}$$

from Lemma 3.1 in [10]. Hence, the operator R_k is monotone.

A pair $(u_k,\gamma_k)\in S_k\times L_k$ is called a "*subsolution*" ("*supersolution*"), if the condition of compatibility (2.2a) is valid and $(u_k,\gamma_k) \leqq R_k(u_k,\gamma_k)$ (resp. $(u_k,\gamma_k)\geqq R_k(u_k,\gamma_k)$). Starting with a non-negative subsolution and a (greater) supersolution, the iteration with R_k gives two monotone sequences, which converge to (possibly different) fixed points of R_k [2]. The following lemma yields that these fixed points are solutions of (2.2).

<u>Lemma 3.1</u>

Let $(u_k,\gamma_k)\in S_k\times L_k$ satisfy the condition of compatibility (2.2a) and let $\underline{d}_k=\underline{b}_k-A_k\underline{u}_k-E_k\underline{\gamma}_k$ denote the defect. Then we have

a)

$$(u_k^i,\gamma_k^i)\geqq(R_k(u_k,\gamma_k))^i \Leftrightarrow \begin{cases} d_k^i\leqq 0 & \text{if } p_i\in D_k, \\ (1-\gamma_k^i)\,d_k^i\leqq 0 & \text{if } p_i\in P_k^o, \\ u_k^i\geqq u^o(p_i) & \text{if } p_i\in P_k^+, \\ \text{and } \gamma_k^i=1 & \text{if } e_{ii}^k=0. \end{cases}$$

b)

$$(u_k^i,\gamma_k^i)\leqq(R_k(u_k,\gamma_k))^i \Leftrightarrow \begin{cases} d^i\geqq 0 & \text{if } p_i\in D_k\cup P_k^o \text{ and } e_{ii}^k\neq 0, \\ u_k^i d_k^i\geqq 0 & \text{if } p_i\in D_k\cup P_k^o \text{ and } e_{ii}^k=0, \\ u_k^i\leqq u^o(p_i) & \text{if } p_i\in P_k^+\cup P_k^o. \end{cases}$$

c) Moreover, if $\underline{b}_k=\underline{0}$, we have

$$(u_k^i,\gamma_k^i)=(R_k(u_k,\gamma_k))^i \Leftrightarrow \begin{cases} d_k^i=0 & \text{if } p_i\in D_k, \\ d_k^i\geqq 0,\ (1-\gamma_k^i)\,d_k^i=0 & \text{if } p_i\in P_k^o, \\ u_k^i=u^o(p_i) & \text{if } p_i\in P_k^o\cup P_k^+, \\ \text{and } \gamma_k^i=1 \quad \text{if } e_{ii}^k=0. \end{cases}$$

The proof of this lemma is straightforward [4].

The Gauß-Seidel version G_k of (3.1) has the same fixed points as R_k [9, Theorem 3.2]. If (u_k,γ_k) is subsolution (supersolution), $G_k(u_k,\gamma_k)$ has the same property [9]. We have found the Gauß-Seidel relaxation to be about twice as fast as the Jacobi relaxation. (See also the numerical results in[9].) The efficiency of both relaxation methods deteriorates quickly with a growing number of grid points. The application of multigrid techniques seems favourable.

The main idea of our two-level algorithm MGDAM is the following:

For fixed γ_k, the first component R_k^u of the operator (3.1) is equal to a Jacobi relaxation with projection as used when solving obstacle problems. Thus to a given approximation $(u_k, \gamma_k) \in S_k \times L_k$ the following obstacle problem is associated (provided $\underline{b}_k = \underline{0}$):

$$\text{Find } v_k \in V_k(u_k,\gamma_k) := \{w_k \in S_k, w_k^i \geqq 0, (1-\gamma_k^i)(w_k^i - u_k^i) = 0 \text{ for } p_i \in P_k,$$
$$w_k^i = u_k^i \quad \text{for } p_i \in P_k^0 \cup P_k^+\},$$

such that (3.3)

$$(\underline{w}_k - \underline{v}_k)^T (A_k \underline{v}_k + E_k \underline{\gamma}_k) \geqq 0$$

holds for every $w_k \in V_k(u_k, \gamma_k)$.

The auxiliary problem (3.3) is considered only in the points of the saturated region $\{\gamma_k = 1\}$ in order to preserve the condition of compatibility (2.2a). Efficient multigrid solvers for obstacle problems are known [6],[8] . We follow the approach in [6] and control the condition of non-negativity in the points of the coarse grid only. The interpolation and restriction operators are determined by the f.e. approach in a natural way. The algorithm MGDAM calculates an increasing sequence of subsolutions and a decreasing sequence of (greater) supersolutions, i.e. it is of the monotonic type.

ALGORITHM MGDAM (one iteration step)

Given a subsolution (u_k^-, γ_k^-) and a supersolution (u_k^+, γ_k^+).

a) *Relaxation.* Perform ν iteration steps

$$(u_k^-, \gamma_k^-) \leftarrow G_k(u_k^-, \gamma_k^-), \quad (u_k^+, \gamma_k^+) \leftarrow G_k(u_k^+, \gamma_k^+)$$

with the Gauß-Seidel version G_k of the operator R_k, defined by (3.1) with $\underline{b}_k = \underline{0}$.

b) *Correction of* u_k. For (u_k^-, γ_k^-) and (u_k^+, γ_k^+) compute an approximate solution v_k^- resp. v_k^+ of the corresponding auxiliary problem (3.3) by one two-level cycle. Put

$$u_k^- \leftarrow \lambda v_k^- + (1-\lambda) u_k^-, \quad u_k^+ \leftarrow \rho v_k^+ + (1-\rho) u_k^+,$$

where $\lambda, \rho \in [0,1]$ are maximal such that the property of subsolution resp. supersolution is preserved.

Lemma 3.1 yields the monotonicity of the calculated sequences. Therefore the convergence of the algorithm is an immediate consequence of the following:

Lemma 3.2.

Suppose that the condition of compatibility (2.2a) is valid for $(u_k^-, \gamma_k^-) \in S_k \times L_k$. Furthermore, let $(u_k^+, \gamma_k^+) \in S_k \times L_k$ satisfy the relations

$$u_k^+(p) \geqq 0\ , \quad \gamma_k^-(p) \leqq \gamma_k^+(p) \leqq 1 \qquad \forall p \in P_k,$$

$$u_k^-(p) \leqq u_k^+(p) \qquad \forall p \in P_k^o \cup P_k^+.$$

Then the inequalities

$$u_k^{-i}(A_k \underline{u}_k^- + E_k \underline{\gamma}_k^-)^i \leqq 0, \quad (A_k \underline{u}_k^+ + E_k \underline{\gamma}_k^+)^i \geqq 0 \quad \forall p_i \in D_k$$

imply $u_k^- \leqq u_k^+$.

Proof: The property as stated follows from the relation

$$(\underline{u}_k^+ - \underline{u}_k^-)_-^T [A_k(\underline{u}_k^+ - \underline{u}_k^-) + E_k(\underline{\gamma}_k^+ - \underline{\gamma}_k^-)] \geqq 0,$$

where $\quad (\underline{u}_k^+ - \underline{u}_k^-)_-^i := -\min\{0, (\underline{u}_k^+ - \underline{u}_k^-)^i\}.$ □

The auxiliary problems (3.3) are approximatively solved by one two-level cycle. Independently, Marini and Pietra [9] used the same auxiliary problem. They, however, solved it by classical relaxation methods.

4. A MULTI-LEVEL ALGORITHM WITH APPROXIMATION OF THE FULL PROBLEM ON THE COARSE GRIDS

In most of the physical situations, the algorithm MGDAM described above turned out to be significantly faster than the Gauß-Seidel relaxation. But MGDAM has not yet the typical multigrid property that the speed of convergence is independent of the mesh size. This is due to the poor approximation of the function γ. Therefore, we introduce a pure multigrid algorithm, called "FASDAM", which uses the full approximation scheme (FAS) [5].

The application of the FAS technique to the discrete dam problem with right hand side $\underline{b}_k$ leads to the following coarse grid problem:

Find $u_{k-1} \in M_{k-1}(u^o)$ and $\gamma_{k-1} \in L_{k-1}$, such that

$$u_{k-1}^i \geqq 0, \ 0 \leqq \gamma_{k-1}^i \leqq 1, \ u_{k-1}^i(1-\underline{\gamma}_{k-1}^i) = 0 \quad \forall p_i \in P_{k-1},$$

$$(A_{k-1}\underline{u}_{k-1}+E_{k-1}\underline{\gamma}_{k-1}-\underline{b}_{k-1})^i \begin{cases} = 0 & \text{for } p_i \in D_{k-1}, \\ \leq 0 & \text{for } p_i \in P^o_{k-1} \end{cases} \tag{4.1}$$

holds, where

$$\underline{b}_{k-1} := A_{k-1}\tilde{I}^{k-1}_k \underline{u}_k + E_{k-1}\tilde{I}^{k-1}_k \underline{\gamma}_k + I^{k-1}_k(\underline{b}_k - A_k\underline{u}_k - E_k\underline{\gamma}_k).$$

We have chosen the restriction operators

$$(\tilde{I}^{k-1}_k \underline{v}_k)^i := v^i_k \quad \text{for } p_i \in P_{k-1} \text{ (injection)}, \tag{4.2}$$

$$(I^{k-1}_k \underline{d}_k)^i := \begin{cases} (I^{k\,T}_{k-1}\underline{d}_k)^i & \text{for } p_i \in P_{k-1} \cap \Omega, \\ 3d^i_k & \text{otherwise.} \end{cases} \tag{4.3}$$

In the f.e.framework, the natural choice for the "residual weighting" operator I^{k-1}_k is the transpose of the natural interpolation matrix I^k_{k-1}. But using this operator and starting with an exact discrete solution, the algorithm does not become stationary. The reason is the bad handling of the conditions of complementarity $((1-\gamma^i_k)d^i_k \leq 0,\ u^i_k d^i_k \geq 0)$ occuring in Lemma 3.1. Solving obstacle problems, Brandt and Cryer studied a similar situation [6, Lemma 5.1].

The modification of the natural residual weighting operator given in (4.3) is sufficient to preserve the stationary points of the relaxation method:

Lemma 4.1

Let $(u^*_k,\gamma^*_k) \in S_k \times L_k$ be a fixed point of R_k with $\gamma^*_k \geq 0$. Then the injection $(u^*_{k-1},\gamma^*_{k-1}) := \sum_{i=1}^{n_{k-1}} (u^{*i}_k \varphi^i_{k-1}, \gamma^{*i}_k \chi^i_{k-1})$ is a fixed point of R_{k-1}.

A detailed proof of this lemma is given in[4]. It exploits the definitions(3.1),(4.1)-(4.3) and Lemma 3.1.

We did not choose the injection for residual weighting in all points of the grid, because this operator is not very robust ; together with a badly chosen multigrid strategy it leads to divergence. Against that, the algorithm FASDAM was always very efficient when we used the restriction (4.3).

ALGORITHM FASDAM (one cycle of the twogrid version)

a) *Smoothing*. Perform ν relaxation steps

$$(u_k,\gamma_k) \leftarrow G_k(u_k,\gamma_k)$$

with the Gauß-Seidel operator G_k defined in MGDAM.

b) *Coarse grid correction*. Determine the corresponding relaxation operator G_{k-1} associated with the coarse grid problem (4.1). Starting with

$$(\underline{u}_{k-1}, \underline{\gamma}_{k-1}) \leftarrow (\tilde{I}_k^{k-1} \underline{u}_k, \tilde{I}_k^{k-1} \underline{\gamma}_k)$$

perform μ iteration steps

$$(u_{k-1}, \gamma_{k-1}) \leftarrow G_{k-1}(u_{k-1}, \gamma_{k-1}).$$

Compute the natural interpolation of the calculated differences

$$\underline{v}_k \leftarrow I_{k-1}^k (\underline{u}_{k-1} - \tilde{I}_k^{k-1} \underline{u}_k), \quad \underline{\delta}_k \leftarrow I_{k-1}^k (\underline{\gamma}_{k-1} - \tilde{I}_k^{k-1} \underline{\gamma}_k)$$

and put

$$\underline{u}_k \leftarrow \max\{\underline{0}, \underline{u}_k + \underline{v}_k\}, \underline{\gamma}_k \leftarrow \min\{\underline{1}, \max\{\underline{0}, \underline{\gamma}_k + \underline{\delta}_k\}\}.$$

The extension of this algorithm in order to get a complete multigrid algorithm is obvious.

5. NUMERICAL RESULTS

We have compared our multi-level algorithms with the pure Gauß-Seidel iteration and with the Incorporated SORP in several physical situations. For this purpose, we have implemented full multigrid versions of MGDAM and FASDAM. Of course an (interpolated) Ω_{k-1}-subsolution (Ω_{k-1}-supersolution) is not a priori an Ω_k-subsolution (Ω_k-supersolution). But a coarse grid subsolution (supersolution) is transformed into a fine grid subsolution (supersolution) by a few relaxation steps provided that the accuracy was not too good. Our programs make use of the adaptive multigrid strategy proposed by Brandt [5],[6] . That is, the numbers of relaxation sweeps and coarse grid corrections are determined by some internal checks. In most of the cases, a V-cycle or a combination of V-cycle and W-cycle is performed, the numbers ν of relaxation sweeps on the finer grids vary between 2 and 5. The value of μ is strongly dependent on the given data of the problem.

All numerical computations were performed on the Control Data 175 in Bochum. The programs were written in FORTRAN 77 and compiled with optimization level 2.

Our first test problem is a homogeneous trapezoidal dam.

The arrows in the following plot describe the velocity of the porous flow. The orthogonal continuous lines are the level lines of the piezometric head $u_k(\underline{x}) + \underline{x} \cdot \underline{e}$. The small vertical lines indicate the unsaturated region $\{0 < \gamma_k < 1\}$. Furthermore, a continuous approximation of the zero line of $u_k + \gamma_k$ is plotted.

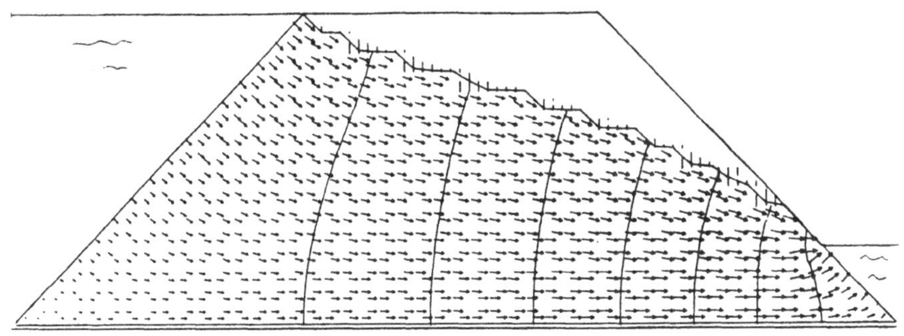

Fig. 3 Problem 1 discretized with 561 nodes

The numerical results are summarized in Table 1 and Table 2.

Table 1: CPU-time [s] solving Problem 1 upto an accuracy of 10^{-2} (for the maximumnorm of the vectors $\underline{u}_k+\underline{\gamma}_k$)

number of nodes (of grids) / method	153(2)	561(3)	2145(4)
Gauß-Seidel relaxation	2.58	38.29	543.87
MGDAM	1.12	8.84	87.96
Incorporated SORP	0.77	5.59	47.03
FASDAM	0.18	0.52	1.51

Table 2: Mean convergence rates of FASDAM solving Problem 1 (first 100 work units [WU]).

number of nodes (of grids)	153(2)	561 (3)	2145(4)
Φreduction of $\lVert defect \rVert_{\ell_2}$/cycle	0.207	0.266	0.256
ϕreduction of $\lVert defect \rVert_{\ell_2}$/WU	0.9	0.876	0.863

The multi-level algorithms turned out to be significantly faster than the comparable relaxation methods.

Next we have considered two inhomogeneous versions of Problem 1. We chose the permeability 1 in the left and 5 in the right half of the dam and vice versa. When we solved these two problems, discretized with 561 nodes, the CPU-time needed by FASDAM varied from 0.58s to 0.50 s. Against that, the CPU-time of the other algorithms changed by nearly 100%.

The following problems are already known from [2].

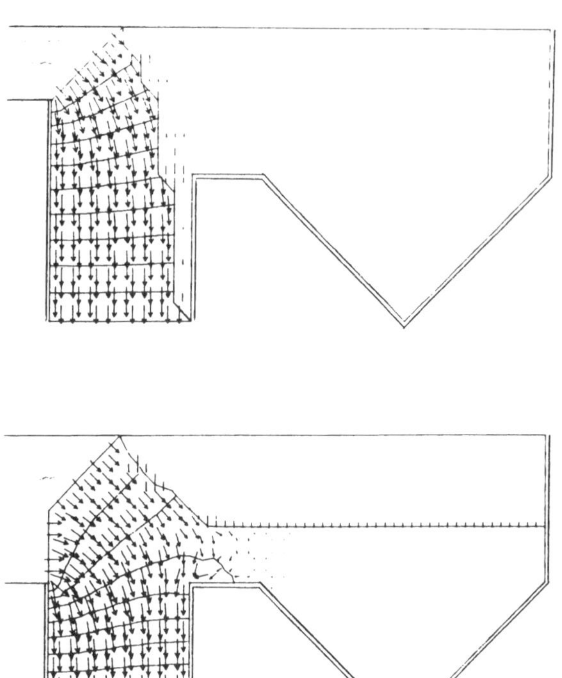

Fig. 4 Problem 2 and Problem 3 discretized with 387 nodes

In Problem 2, every water level in the lower right part of the porous medium gives another possible solution. But if we change the data a little bit, as in Problem 3, the solution becomes unique.

In Problem 4, the permeability is 1 in the upper and 8 in the lower half of the dam. The solution has a wide unsaturated zone between two disconnected saturated regions.

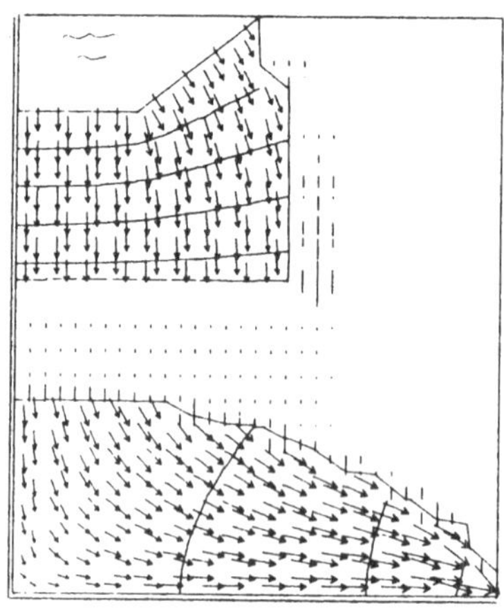

Fig. 5 Problem 4 discretized with 399 nodes

with the matrix U.

2. FORMULATION OF THE EIGENV

In this chapter, L denotes an nxn-m
sumed to be large. We want to compute k
vectors), where k<<n.

Lemma 1 For any nxn matrix L there ex
an upper triangular kxk matrix A satisf

$$L\,U - U\,A = 0$$

$$U^{*}U \quad = I \quad \text{(I is the}$$

The diagonal entries of A represent k

Proof. There exist a unitary nxn matri
lar nxn matrix R with LQ-QR=O. Let A b

($1 \leq i, j \leq k$) and let U consist of the fir
A and U satisfy Eqs (1a,b).

Note 2 (a) The matrices U and A satisf
means unique. (b) If the k eigenvalues

ones, the solution U, A is isolated. (
A is diagonal and the columns of U con
L. (d) If A is diagonalisable, i.e., if
AT=TD (D diagonal), then the columns o
tors of L belonging to the eigenvalues

3. NEWTON'S ITERAT

Let U and A be a solution of syst
denote given approximations. We define

$$\delta U := \tilde{U}-U, \qquad \delta A := \tilde{A}-A.$$

Replacing U and A in (1a,b) by $\tilde{U}-\delta U$ ar
matrix equations

$$L\,\delta U - \delta U\,\tilde{A} - \tilde{U}\,\delta A = F -$$

$$\delta U^{*}\tilde{U} + \tilde{U}^{*}\delta U = G + \delta U^{*}\delta U,$$

where F and G are the defects

$$F := L\,\tilde{U} - \tilde{U}\,\tilde{A}, \qquad G :=$$

The matrix equations (3a) represent n
unknown entries of δU. Since the unkn
angular, the number of its nonzero el

The multigrid algorithm of the monotonic type MGDAM turned out to be unsuitable for solving Problem 4. Besides this, the performance of the algorithms is always similar as observed in Problem 1:

Calculating the plotted solutions, MGDAM required between 16% and 43% of the CPU-time needed by the Gauß-Seidel relaxation and FASDAM between 3% and 13% of the time needed by the Incorporated SORP. The ratios become more favourable for the multigrid algorithms with a growing number of grid points.

For more details, we refer to [4], where the treatment of an actual dam of the river Rhine is also reported.

R E F E R E N C E S

[1] Alt, H.W.:" Strömungen durch inhomogene poröse Medien mit freiem Rand", J. Feine Angew.Math. 305(1979)pp.89-115.

[2] Alt, H.W.:"Numerical solution of steady-state porous flow free boundary problems", Numer.Math. 36(1980) pp.73-98.

[3] Baiocchi,C.:" Su un problema di frontiera libera connesso a questione di idraulica", Ann.Mat.Pura Appl.(4)92 (1972) pp. 107-127.

[4] Bollrath,C.:"Zwei Mehrgitterverfahren zur numerischen Berechnung von stationären Strömungen durch poröse Medien mit freiem Rand", Dissertation,Ruhr-Universität Bochum (submitted).

[5] Brandt,A.:"Multi-level adaptive solutions to boundary value problems", Math.Comp. 31(1977) pp. 333-390.

[6] Brandt,A., Cryer,C.W.:"Multigrid algorithms for the solution of linear complementarity problems arising from free boundary problems", SIAM J.Sci.Stat.Comput.4 (1983) pp. 655-684.

[7] Brezis,H., Kinderlehrer,D., Stampacchia,G.:" Sur une nouvelle formulation du problème de l'écoulement à travers une digue", C.R. Acad.Sci. Paris 287(1978) pp. 711-714.

[8] Mandel,J.:" A fast iterative method for large, sparse, symmetric, positive definite linear complementary systems", Appl.Math.Optim. 11(1984) pp. 77-95.

[9] Marini,L.D., Pietra,P.:"Fixed-point algorithm for stationary flow in porous media", Instituto di Analisi Numerica del C.N.R., Pavia.Preprint (1983).

[10] Pietra,P.:"An up-wind method for a filtration problem", RAIRO, Anal.Numér.(4) 16(1982) pp. 483-481.

MULTI-GRID EIGENVALU

W. Hackbus

Institut für Informatik und I

Christian-Albrechts-U

Olshausenstr. 40, D-2300

SUMMARY

A multi-grid iteration is desc
multaneously k eigenvalues and the
trix may be unsymmetric or even not
rithm is based on a Newton iteratio
matrix of the Schur normal form.

1. INTRODUCT

There exist several approaches
blems $Lu=\lambda u$. For instance, an invers
by Bank [1]. A direct multi-grid ap
busch [5]. This algorithm produces
necessarily the smallest one) and t
application is given by Hackbusch-H
any eigenvalue problem can be formu
with an isolated solution (cf Brandt
Often the computation is more stabl
and eigenvectors is calculated simu
Brandt-McCormick-Ruge [2] is of thi
jection and can be applied only to
A simultaneous Newton iteratio
cable to general matrices is descri
termines a subspace span$\{u_1,\dots,u_k\}$,
of a matrix U. The unknown matrix U
$V^TU=I$ (identity), where V is a give
nonlinear equation to be solved tak

$$F(U) := LU - U(V^TLU)$$

Each Newton step requires the soluti
the kind $A^*\delta U-\delta U^*B=C$ (A,B,C given).
matrix V^TLU coincide with k eigenva
eigenvalues cannot be read directly
a full matrix.
In the following approach we r
upper triangular matrix A, which ha

also the number of relevant equations in (3b), since both sides of Eq (3b) are symmetric.

The existence of solutions δU, δA is ensured by Lemma 1. The Newton iteration can be obtained for Eqs (3a-c) by omitting the quadratic terms.

Newton iteration: $\tilde{U} \mapsto \tilde{U} - \delta U$, $\tilde{A} \mapsto \tilde{A} - \delta A$, where

$$L\,\delta U - \delta U\,\tilde{A} - \tilde{U}\,\delta A = F, \tag{4a}$$

$$\delta U^*\tilde{U} + \tilde{U}^*\delta U = G. \quad (F,G \text{ from } (3c)) \tag{4b}$$

4. FORMULATION OF SYSTEM (4) AS A STAGGERED SYSTEM

In the following, we assume that

the eigenvalues $a_{11},\dots,a_{kk}$ are simple, (5a)

L and the eigenvalues $a_{11},\dots,a_{kk}$ are real. (5b)

Note 3 By (5b) the matrices U and A (accordingly also $\tilde{U},\tilde{A},\delta U,\delta A$) can be assumed to be real. Hence, $U^*=U^T$, $\delta U^*=\delta U^T$ etc.

The first column of the matrix equation (4a) reads as

$$L\,\delta u_1 - \delta u_1\tilde{a}_{11} - \tilde{u}_1\,\delta a_{11} = f_1, \tag{6a}$$

where δu_j, $\tilde{u}_j$, f_j are the jth columns of δU, $\tilde{U}$, F, respectively. $\tilde{a}_{ij}$ and δa_{ij} are the entries of $\tilde{A}$ and δA. The (1,1) entries of the matrices in Eq (4b) yield the equation

$$\delta u_1^T\,\tilde{u}_1 + \tilde{u}_1^T\delta u_1 = g_{11}. \tag{6b}$$

Eqs (6a,b) can be written as

$$\begin{bmatrix} L-\tilde{a}_{11}I & -\tilde{u}_1 \\ -\tilde{u}_1^T & 0 \end{bmatrix} \begin{bmatrix} \delta u_1 \\ \delta a_{11} \end{bmatrix} = \begin{bmatrix} f_1 \\ -g_{11}/2 \end{bmatrix}, \tag{7a}$$

which is a linear equation of standard form. From (7a) one can compute δu_1 and δa_{11} (i.e. the first columns of δU and δA). Assume that the first j-1 columns of δU and δA are already determined. The jth columns in Eqs (5a,b) yield

$$L\,\delta u_j - \delta u_1\tilde{a}_{1j} - \dots - \delta u_j\tilde{a}_{jj} - \tilde{u}_1\delta a_{1j} - \dots - \tilde{u}_j\;a_{jj} = f_j,$$

$$\delta u_i^T\tilde{u}_j + \tilde{u}_i^T\delta u_j = g_{ij}\ (i=1,\dots,j).$$

These equations are equivalent to

$$\begin{pmatrix} L-\tilde{a}_{jj}I & -\tilde{u}_1 & -\tilde{u}_2 & \cdots & -\tilde{u}_j \\ -\tilde{u}_1^T & & & & \\ -\tilde{u}_2^T & & & 0 & \\ \vdots & & & & \\ \tilde{u}_j^T & & & & \end{pmatrix} \begin{pmatrix} \delta u_j \\ \delta a_{1j} \\ \delta a_{2j} \\ \vdots \\ \delta a_{jj} \end{pmatrix} = \begin{pmatrix} f_j + \sum_{i=1}^{j-1} \tilde{a}_{ij}\delta u_i \\ -g_{1j}+\delta u_1^T\tilde{u}_j \\ -g_{2j}+\delta u_2^T\tilde{u}_j \\ \vdots \\ -g_{jj}/2 \end{pmatrix} \tag{7b}$$

The right-hand side involves only previously computed quantities.

Note 4 The matrix equations (4a,b) are equivalent to the staggered system of the problems (7b) for $j=1,\ldots,k$.

If $\tilde{u}_i$ $(1\leq i\leq k)$ are sufficiently close to u_i and if a_{ii} $(1\leq i\leq k)$ are different from the other eigenvalues, the matrices of System (7b) are regular. This proves

Lemma 5 Under conditions (5a,b) the solution U, A of Eqs (1a,b) is isolated and the Newton iteration is locally quadratically convergent.

Numerical example. Consider the eigenvalue problem

$$-u''(x) - 10u'(x) = \lambda u(x) \text{ for } x\in(0,1),\ u(0)=u(1)=0. \tag{8}$$

The standard difference approximation with a one-sided difference $[u(x)-u(x-h)]/h$ for $u'(x)$ yields an unsymmetric matrix. For $h=1/8$, $k=3$ we use the starting matrices

$$\tilde{U}_{ij} = \frac{1}{2}\sin(j\pi ih) \quad (1\leq j\leq k=3),\ \tilde{A} = 0,$$

since the columns of $\tilde{U}$ are the eigenvectors of $-u''=\lambda u$. The 3×3 matrices A, $\tilde{U}^T\tilde{U}$ and the norms of the 3 columns of the defect $F:=L\tilde{U}-\tilde{U}\tilde{A}$ are shown below.

	matrix $\tilde{A}$			matrix $\tilde{U}^T\tilde{U}$			norms of the columns of $F=L\tilde{U}-\tilde{U}\tilde{A}$
starting iterate	0	0	0	1	0	0	30.9
	0	0	0	0	1	0	78.2
	0	0	0	0	0	1	143.5
first iterate	25.0	19.9	-6.8	1.13	0.03	-0.09	9.1
	0	82.1	24.7	0.03	1.21	0.04	39.8
	0	0	151.6	-0.09	0.04	1.15	63.2

Table continued

	matrix $\tilde{A}$			matrix $\tilde{U}^T\tilde{U}$			norms of the columns of $F=L\tilde{U}-\tilde{U}\tilde{A}$
second iterate	29.4	56.1	15.2	1.034	0.008	-0.069	0.81
	0	81.5	91.6	0.008	1.193	0.053	6.69
	0	0	157.3	-0.069	0.053	1.419	30.68
third iterate	30.56	54.48	18.51	1.0007	-0.0012	-0.0011	0.0305
	0	72.56	80.55	-0.0012	1.0097	0.0045	0.8603
	0	0	137.14	-0.0011	0.0045	1.0370	4.2962
fourth iterate	30.6151	54.0081	18.8473	1.0000	-0.0000	-0.0000	0.0000
	0	72.2392	79.1999	-0.0000	1.0001	-0.0001	0.0022
	0	0	134.6194	-0.0000	-0.0001	1.0007	0.0665
fifth iterate	30.6151	54.0060	18.8370	1.0000	-0.0000	0.0000	0.0000
	0	72.2355	79.1290	-0.0000	1.0000	0.0000	0.0000
	0	0	134.5247	0.0000	0.0000	1.0000	0.0001

The exact eigenvalues are $\lambda_\nu=16[1+24\sin^2(\nu\pi/16)]$:

$\lambda_1=30.61512975$, $\lambda_2=72.23549805$, $\lambda_3=134.5247809$.

5. MULTI-GRID SOLUTION OF EQS (4a,b)

Let $L_0,\ldots,L_l$ be a sequence of discretisations corresponding to refining grids. For all levels the matrices A_l have constant size kxk. We assume that for the levels $1\le k\le l-1$ the matrices $\tilde{U}_k$ and $\tilde{A}_k$ are already approximated. A (column-wise) interpolation of $\tilde{U}_{l-1}$ serves as the starting guess $\tilde{U}_l$, while $\tilde{A}_l:=\tilde{A}_{l-1}$ is the initial value at level l. These approximations are to be improved by means of the Newton iteration, which requires the solution of Eqs (4a,b) at level l:

$$L_l\delta U_l - \delta U_l\tilde{A}_l - \tilde{U}_l\delta A_l = F_l, \tag{9a}$$

$$\delta U_l^T\tilde{U}_l + \tilde{U}_l^T\delta U_l = G_l. \tag{9b}$$

The multi-grid approach can follow the structure of the equations (7b). A smoothing process has to be applied to

$$(L_l-\tilde{a}_{jj}I)\delta u_j=f_j+\sum_{i=1}^{j-1}\delta u_i\tilde{a}_{ij}+\sum_{i=1}^{j}\tilde{u}_i\delta a_{ij}, \tag{10}$$

where $\tilde{u}_i$, δu_i, f_i are the ith columns of $\tilde{U}_l$, δU_l, F_l and where $\tilde{a}_{ij}$, δa_{ij} are the entries of A_l, δA_l.

Since Eq (10) is (almost) singular, the smoothing process yields a result, which does not satisfy the last equation of (7b) and which may contain large errors with respect to u_i-components with $a_{ii} \approx a_{jj}$. The correction of these u_i-components and of δA_l could be expected from the coarse-grid correction. However, it is not difficult to improve these quantities also at the level l (e.g. before smoothing is performed).

Let $\widetilde{\delta U}=\widetilde{\delta U}_l$ be an approximation of the solution δU_l of Eqs (9a,b).
Assume that the errors

$$\tilde{F} := L\,\tilde{U} - \tilde{U}\,\tilde{A}, \quad \delta F := L\,\widetilde{\delta U} - \widetilde{\delta U}\,\tilde{A} - \tilde{U}\,\widetilde{\delta A} - F \qquad (11a)$$

of $\tilde{A}=\tilde{A}_l$, $\tilde{U}=\tilde{U}_l$ and of $\widetilde{\delta U}=\widetilde{\delta U}_l$ lie in the subspace spanned by the columns of $\tilde{U}$, i.e. there exist kxk matrices $\tilde{C}$ and C such that

$$\tilde{F} = \tilde{U}\,\tilde{C}, \qquad \delta F = \tilde{U}\,C.$$

The further residuals are denoted by

$$\tilde{G} := \tilde{U}^T\tilde{U} - I, \quad \delta G := \delta U^T\tilde{U} + \tilde{U}^T\delta U - G. \qquad (11b)$$

In the Newton process F, G coincide with $\tilde{F}$ and $\tilde{G}$. In the multigrid process this is not necessarily true. The solution of Eq (11a) is of the form

$$\delta U = \widetilde{\delta U} - \delta\delta U, \; \delta\delta U = \tilde{U}\,B \;\; \text{(B is a kxk matrix)}, \; \delta A=\widetilde{\delta A}-\delta\delta A. \qquad (11c)$$

B (i.e. $\delta\delta U$) and $\delta\delta A$ have to satisfy $L\tilde{U}B-\tilde{U}B\tilde{A}-\tilde{U}\delta\delta A=\delta F$ and $B^T\tilde{U}^T\tilde{U}+\tilde{U}^T\tilde{U}B=\delta G$.
Replacing $L\tilde{U}$ by $\tilde{U}\tilde{A}+\tilde{F}=\tilde{U}(\tilde{A}+\tilde{C})$ and multiplying by $\tilde{U}^T$ from the left, we transform the first equation into $(I+\tilde{G})[(\tilde{A}+\tilde{C})B-B\tilde{A}-\delta\delta A]=(I+\tilde{G})C$. Together with the second one we obtain the two matrix equations

$$(\tilde{A} + \tilde{C})\,B - B\,\tilde{A} - \delta\delta A = C, \qquad (12a)$$

$$B^T(I+\tilde{G}) + (I + \tilde{G})\,B = \delta G \qquad (12b)$$

for the unknown matrices B and $\delta\delta A$. Again these quantities can be computed by solving a staggered system of usual equations. For $j=1,2,\ldots,k$ one has to determine the jth column b_j of B as the solution of the following k equations:

$$b_{ij} + \sum_{m=1}^{k} \tilde{g}_{im} b_{mj} = \delta g_{ij} - b_{ji} - \sum_{m=1}^{k} b_{mi}\tilde{g}_{mj} \qquad (1 \leq i<j), \qquad (13a)$$

$$b_{ij} + \sum_{m=1}^{k} \tilde{g}_{jm} b_{mj} = \delta g_{jj}/2 \qquad (i=j), \qquad (13b)$$

$$-\tilde{a}_{jj}b_{ij} + \sum_{m=1}^{k} (\tilde{a}_{im}+\tilde{c}_{im})\,b_{mj} = c_{ij} + \sum_{m=1}^{j-1} b_{im}\tilde{a}_{mj} \qquad (j<i\leq k). \qquad (13c)$$

The upper triangular matrix $\delta\delta A$ is explicitly given by

$$\delta\delta a_{ij} = -c_{ij} - \sum_{m=1}^{j} b_{im}\tilde{a}_{mj} + \sum_{m=1}^{k} (\tilde{a}_{im}+\tilde{c}_{im}) b_{mj} \quad (1\le i\le j). \tag{13d}$$

Having computed b_j and $\delta\delta a_{ij}$, we can correct the jth columns of δU and δA by

$$\tilde{\delta u}_j \longmapsto \delta u_j := \tilde{\delta u}_j - \sum_{i=1}^{k} b_{ij}\tilde{u}_i, \quad \tilde{\delta a}_{ij} \mapsto \delta a_{ij} := \tilde{\delta a}_{ij} - \delta\delta a_{ij} \tag{13e}$$

(cf. Eqs (11c)).

<u>Note 6</u> The correction described above require the computation of the matrices $\tilde{G}$, δG, and

$$\tilde{C} := (I+\tilde{G})^{-1}\tilde{U}^T\tilde{F}, \qquad C := (I+\tilde{G})^{-1}\tilde{U}^T\delta F. \tag{14}$$

Under condition (5a,b), System (13a-c) is soluble, if $\tilde{G}$, $\tilde{C}$, and $A-\tilde{A}$ are sufficiently small.

<u>Multi-grid process for solving Eqs (9a,b)</u>. Start with $\delta U_l := 0$, $\delta A_l := 0$ and repeat the following iteration.

correct δU and δA according to (13a-e); (15a)

for j:= 1 **to** k **do** smooth Eq (10); (15b)

compute the restricted defects

$$F_{l-1} := r[L_l\,\delta U_l - \delta U_l\tilde{A}_l - \tilde{U}_l\delta A_l - F_l]; \quad G_{l-1} := \delta U_l^T\tilde{U}_l + \tilde{U}_l^T\delta U_l - G_l; \tag{15c}$$

solve the coarse-grid equations

$$L_{l-1}\delta U_{l-1} - \delta U_{l-1}\tilde{A}_{l-1} - \tilde{U}_{l-1}\delta A_{l-1} = F_{l-1}, \tag{15d$_1$}$$

$$\delta U_{l-1}^T\tilde{U}_{l-1} + \tilde{U}_{l-1}^T\delta U_{l-1} = G_{l-1} \tag{15d$_2$}$$

by γ (γ=1 or 2) multi-grid iterations at level l-1;

correct by $\delta U_l := \delta U_l - p\delta U_{l-1}$; $\delta A_l := \delta A_l - \delta A_{l-1}$; (15e)

r and p are the usual restriction and prolongation operators, which have to be applied to each of the k columns of the matrix.

The multi-grid convergence of Algorithm (15) can be seen from the following numbers. The eigenvalue problem $u''=\lambda u$ in (0,1) with u(0)=u(1)=0 is discretised by $h_o = 1/4$, $h_1 = 1/8$. The number of simultaneously computed eigenvalues is k=2. During the first step of the Newton iteration of Section 3 we apply 4 iterations of Algorithm (15). The smoothing is performed by 2 Gauß-Seidel steps. Let $(d_1,d_2) := D := L_l\delta U_l - \delta U_l\tilde{A}_l - \tilde{U}_l\delta A_l - F_l$ be the defect of the approximations δU_l, δA_l. The norms of d_1 and d_2 are shown below.

number of iterations	0	1	2	3	4
$\|\|d_1\|\|$	9.02	0.89	0.085	0.0058	0.0007
$\|\|d_2\|\|$	27.7	6.19	0.769	0.1720	0.0412

6. MODIFICATIONS AND GENERALISATIONS

The correction (15a) can be omitted or replaced by a simpler one. Furthermore, the correction step (15a) can be performed columnwise in parallel to the smoothing process.

Condition (5a) excludes multiple eigenvalues. In the case of a double eigenvalue $\lambda_1=\lambda_2$ there arise two problems: 1) The solution U, A of Eqs (1a,b) is not isolated. 2) Although the eigenvalues depend continuously of the matrix L, the matrix U may be a discontinuous function of L, when λ passes a double eigenvalue.

To obtain U and A as an isolated solution, we have to add a further equation

$$w^T u_1 = 0. \tag{16}$$

On the other hand, we have to increase the number of unknowns: a_{21} may be a nonzero entry of A. Then the equations (7b) are no longer decoupled for j=1 and j=2. Instead we obtain the system

$$\left[\begin{array}{cc|cccc}
L-\tilde{a}_{11}I & -\tilde{a}_{21}I & -\tilde{u}_1 & 0 & -\tilde{u}_2 & 0 \\
-\tilde{a}_{12}I & L-\tilde{a}_{22}I & 0 & -\tilde{u}_1 & 0 & -\tilde{u}_2 \\
\hline
-\tilde{u}_1^T & 0 & 0 & 0 & 0 & 0 \\
0 & -\tilde{u}_2^T & 0 & 0 & 0 & 0 \\
-\tilde{u}_2^T & -\tilde{u}_1^T & 0 & 0 & 0 & 0 \\
-w^T & 0 & 0 & 0 & 0 & 0
\end{array}\right]
\left[\begin{array}{c}
\delta u_1 \\ \delta u_2 \\ \hline \delta a_{11} \\ \delta a_{12} \\ \delta a_{21} \\ \delta a_{22}
\end{array}\right]
=
\left[\begin{array}{c}
f_1 \\ f_2 \\ \hline -g_{11}/2 \\ -g_{22}/2 \\ -g_{12} \\ w^T\tilde{u}_1
\end{array}\right]$$

Similarly, a modification is needed if pairs of complex conjugate eigenvalues appear.

REFERENCES

[1] BANK, R.A.: Analysis of a multilevel inverse iteration procedure for eigenvalue problems. SIAM J. Numer. Anal. 19 (1982) 886-898.

[2] BRANDT, A., McCORMICK, S., and J. RUGE: Multigrid methods for differential eigenproblems. SIAM J. Sci. Statist. Comput. 4 (1983) 244-260.

[3] CHATELIN, F.: Simultaneous Newton's iteration for the Eigenproblem. Computing, Suppl. 5 (1984) 67-74.

[4] CHATELIN, F.: Spectral approximation of linear operators. Academic Press, New York 1983.

[5] HACKBUSCH, W.: On the computation of approximate eigenvalues and eigenfunctions of elliptic operators by means of a multi-grid method. SIAM J. Numer. Anal. 16 (1979) 201-215.

[6] HACKBUSCH, W.: Multi-grid solutions to linear and nonlinear eigenvalue problems for integral and differential equations. Rostock Math. Colloq. 25 (1984) 79-98.

[7] HACKBUSCH, W.: Multi-Grid Methods. Springer-Verlag, Berlin 1985 (to appear)

[8] HACKBUSCH, W. and G. Hofmann: Results of the eigenvalue problem for the plate equation. Z. Angew. Math. Phys. 31 (1980) 730-739.

[9] HOFMANN, G.: Analysis eines Mehrgitterverfahrens zur Berechnung von Eigenwerten elliptischer Differentialoperatoren. Doctoral thesis, Kiel 1985.

MULTIGRID SOLUTION OF THE STEADY EULER EQUATIONS.

P.W. Hemker and S.P. Spekreijse

CWI, Centre for Mathermatics and Computer Science
P.O. Box 4079, 1009 AB Amsterdam, The Netherlands

SUMMARY

A multigrid (MG) method for the approximation of steady solutions to the full 2-D Euler equations is described. The space discretization is obtained by the finite volume technique and Osher's approximate Riemann-solver. Symmetric Gauss-Seidel relaxation is applied to solve the nonlinear discrete system of equations. A multigrid method, the full approximation scheme, accelerates this iterative process.

In a few two-dimensional testproblems, (subsonic, transsonic and supersonic) the multigrid iteration is applied to an initial estimate that was obtained by means of the FMG-technique (nested iteration). For the discretization on the different levels, a fully consistent sequence of nested discretizations is used. The prolongations and restrictions selected are in agreement with this consistency.

It turns out that the total amount of work required to obtain a solution, that is accurate upto truncation error, corresponds to a small number of nonlinear Gauss-Seidel iterations. In the case of transsonic flow the rate of convergence of the MG-iteration appears independent of N, i.e. the number of cells in the discretization.

1. INTRODUCTION

The Euler equations for compressible inviscid flow in a two dimensional domain Ω,

$$q_t + f_x(q) + g_y(q) = 0, \tag{1.1}$$

form a quasi-linear hyperbolic system of conservation laws. The state of the fluid at a point $(x,y)\in\Omega$ is given by $q(x,y) = (\rho,\rho u,\rho v,E)^T$. The fluxes in the $x-$ (resp. $y-$) direction are

$$f(q) = (\rho u,\rho u^2 + p,\rho uv,u(E+p))^T$$

and

$$g(q) = (\rho v,\rho vu,\rho v^2 + p,v(E+p))^T.$$

For a perfect gas, p and E are related by the equation of state

$$p = (\gamma-1)\,(E-\tfrac{1}{2}\rho(u^2 + v^2)).$$

With (n_1,n_2) a pair of direction cosines, the flux in the (n_1,n_2)- direction is given by $n_1 f + n_2 g$. It is easily verified that the Euler equations are invariant under rotation of the independent variables. This means that, with the change of variables

$$\begin{bmatrix} x' \\ y' \end{bmatrix} = \begin{bmatrix} n_1 & n_2 \\ -n_2 & n_1 \end{bmatrix} \begin{bmatrix} x \\ y \end{bmatrix}$$

we obtain

$$q'_t + f_{x'}(q') + g_{y'}(q') = 0\,,$$

where

$$q' = \begin{bmatrix} 1 & 0 & 0 & 0 \\ 0 & n_1 & n_2 & 0 \\ 0 & -n_2 & n_1 & 0 \\ 0 & 0 & 0 & 1 \end{bmatrix} q.$$

Solutions of (1.1) are not necessarily smooth functions and it makes sense to generalize (1.1) to its weak form:

$$\frac{d}{dt}\iint_{\Omega'} q(x,y)dxdy + \oint_{\partial\Omega'}(n_1f + n_2g)ds = 0\,, \tag{1.2}$$

for all $\Omega' \subset \Omega$.

where (n_1,n_2) is the direction of the outward normal along $\delta\Omega'$. Equation (1.2) allows also non-physical solutions and it can be shown that a physical solution should also satisfy an entropy condition [6,14].

In symbolic form we write (1.1) or (1.2) as

$$q_t + N(q) = 0. \tag{1.3}$$

The steady Euler equations are given by

$$N(q) = 0. \tag{1.4}$$

Here $N:X \rightarrow Y$ is a nonlinear operator, $X \subset [L^2(\Omega)]^4$ is the space of possible fluid states and $Y = [L^2(\Omega)]^4$ is the Banach space of rates of change (of state).

NOTATION

$\Omega \subset \mathbb{R}^2$ the domain of definition for the Euler equations.
Ω_{ij} a cell in the partitioning of Ω.
$\partial\Omega_{ij}$ the boundary of Ω_{ij}.
Ω_{ijk} a neighbouring cell of Ω_{ij}, k=N,S,E,W.
$\Gamma_{ijk} = \bar{\Omega}_{ij} \cap \bar{\Omega}_{ijk}$.
q_{ij} an approximation for the mean state in Ω_{ij}

$$q_{ij}\cdot \text{meas}\,(\Omega_{ij}) = \int_{\Omega_{ij}} q(x,y)dxdy.$$

q_{ijk} an approximation for the mean state in Ω_{ijk}.
$q_h = \{q_{ij} \mid \Omega_{ij} \subset \Omega\}$.

p	pressure.
ρ	density.
u,v	speed of fluid in x,y direction.
E	total energy per unit volume.
c	$= \sqrt{\gamma p / \rho}$ speed of sound.
z	$= \ln(p\,\rho^{-\gamma})$ specific entropy.
γ	ratio of specific heats, $\gamma = 1.4$.

2. DISCRETIZATION

In order to achieve a discretization of a hyperbolic system of conservation laws that has a significant meaning even if the meshsize is coarse, it is important to use a fully conservative method which discretizes the weak form of the equations. A simple but effective technique is found in the finite volume discretization. This technique allows for an irregular partitioning of the domain Ω in a number of disjunct cells. For ease of notation and implementation, we divide the bounded domain Ω in quadrilateral cells Ω_{ij}, such that the result is topological equivalent with a partitioning in regular squares. In each cell the state of the fluid is represented by q_{ij}, the mean value of $q(x,y)$ over Ω_{ij}. The semi-discrete form of the space-discretized equation is

$$\text{meas}\,(\Omega_{ij})\frac{d}{dt}q_{ij} + \oint_{\delta\Omega_{ij}}(n_1f + n_2g)ds = 0, \text{ for all } \Omega_{ij} \subset \Omega, \tag{2.1}$$

where $n_1 f + n_2 g$ denotes the normal flux outward Ω_{ij}. For the steady equation on our mesh this reduces to

$$\sum_{k=N,E,S,W} \int_{\Gamma_{ij}} (n_1 f + n_2 g) ds = 0, \text{ for all } i,j, \tag{2.2}$$

Here $\int_{\Gamma_{ij}} (n_1 f + n_2 g) ds$ denotes the rate of transport of q over the boundary Γ_{ijk} from cell Ω_{ij} to the neighbouring cell Ω_{ijk}, i.e. the flux from Ω_{ij} to Ω_{ijk} multiplied by meas (Γ_{ijk}). In the discretization this flux $n_1 f + n_2 g$ is approximated by the *numerical flux* $f(q_{ij}, q_{ijk})$. In a first order discretization scheme this numerical flux over Γ_{ijk} depends only on the unknown states q_{ij} and q_{ijk}. For a given choice of the numerical flux, the discrete system for the steady equations (2.2) can formally be written

$$N_h(q_h) = 0 . \tag{2.3}$$

This is the discrete system of nonlinear equations of which the solution is required. The operator $N_h : X_h \rightarrow Y_h$ is the nonlinear discrete operator, X_h is the (finite dimensional) linear space of mean states in cells Ω_{ij}, and Y_h is the (finite dimensional) linear space representing the rate of netto transport into the cells Ω_{ij}.

To obtain a good discretization, the selection of a good numerical flux is essential. Such a flux should take into account the fact that, depending on the local characteristics, the flux at Γ_{ijk} is -in a specific way - depending on q_{ij} and q_{ijk}. E.g. in case of a supersonic flow from Ω_{ij} to Ω_{ijk} this flux depends only on q_{ij}.

One way to determine the numerical flux is to consider the flux computation at Γ_{ijk} as a locally one-dimensional problem and to solve the Riemann problem of gasdynamics: compute the flux at Γ_{ijk}, $0 < t \leq t_0$ with at $t = 0$ the initial conditions $q = q_{ij}$ in Ω_{ij} and $q = q_{ijk}$ in Ω_{ijk}. The use of this computed flux as the numerical flux in (2.3) yields the Godunov discretization. A disadvantage is the expensive solution of the Riemann problem at each cell boundary. Several less expensive approximate Riemann solvers have been proposed. These lead to various well known flux (difference) splitting methods [2,8,9,11,12,15,17,18].

Selecting a numerical flux,we should take care that it should (1) yield shocks which satisfy correct jump conditions, (2) find only physically correct shocks, i.e. shocks satisfying an entropy condition, and (3) yield nonlinear stability (a monotone nonlinear system (2.3)).
Further it is an advantage to have a differentiable function $f(q_{ij}, q_{ijk})$ if a Newton-type technique is used in the solution process for (2.3).

A very good numerical flux, that satisfies these desired properties, is generated by Osher's approximate Riemann-solver [8,9]. This numerical flux is based on the Riemann-invariants which relate different states that are connected by simple waves [14]. The good qualities of this numerical flux are often considered to be offset by its relative complexity. However, we found this disadvantage improved by the use of a proper set of dependent variables,viz. u, v, c and z [3].

An additional advantage of the numerical flux computation based on Osher's approximate Riemann solver is the fact that the treatment of boundary conditions, i.e. the computation of the boundary fluxes, can be done in a way that is completely consistent with the computation of fluxes over interior cell walls. For this purpose the equations are considered quasi-one-dimensional, normal to $\delta\Omega$, and a state q_{ijB} at the boundary $\overline{\delta\Omega_{ij} \cap \delta\Omega}$ can be calculated,such that q_{ijB} satisfies the boundary conditions and such that q_{ijB} and q_{ij} are connected by Riemann invariants that correspond to negative (right boundary) or positive (left boundary) eigenvalues [3].

It appears that in the MG-iteration such a consistent treatment of the equation and the boundary conditions makes a special relaxation treatment of the boundary superfluous. (A similar effect is seen when an elliptic equation together with its boundary conditions is consistently dicretized by the finite element method.)

3. SOLUTION METHODS AND LINEARIZATION

Globally there are three ways to solve the nonlinear system (2.3). First, one can start from the semi-discretized form (2.1) and solve the large system of ordinary differential equations

$$D_h \frac{d}{dt} q_h + N_h(q_h) = 0 \qquad (3.1)$$

by some explicit time integrator; D_h is a diagonal matrix. In this way the time-dependent behaviour of the flow is followed, starting from some initial condition. Integration over a long enough time interval may make the solution of (3.1) converge to the solution of (2.3). An advantage is that intermediate values of q_h allow a physical interpretation. If (2.3) has a non-unique solution, the proper choice of an initial condition may select the solution required. Another advantage is that this method needs only evaluations of the operator $N_h(q_h)$. If the time-dependent solution is not wanted, it is a disadvantage that usually many time-steps are needed to obtain a sufficiently converged final solution q_h. Stability conditions may prevent the use of large time steps. Multigrid may provide a technique to accelerate the convergence of this time stepping [4,13].

Secondly, the equation (2.3) may be solved by some implicit time integrator, or -what is closely related- the system (2.3) or (3.1) can be solved by some global linearization. Examples are the application of a Newton-type method or the use of the SER- (Switched Evolution Relaxation) scheme as used by Mulder and Van Leer [7],

$$[D_h + \Delta t \cdot N_h'(q_h^{(n)})]\,(q_h^{(n+1)} - q_h^{(n)}) = -\Delta t \cdot N_h(q_h^{(n)})\,. \qquad (3.2)$$

In these methods large linear systems are to be solved and the construction of the Jacobian matrix

$$N_h'(q_h)$$

is required. For the solution of the linear system (e.g. in eq. 3.2), several techniques are available. Multigrid in its linear form (the Correction Scheme) [1] can be used to accelerate iterative methods for the solution of the linear systems.
In particular Newton's method will give very fast convergence, provided that a sufficiently accurate initial estimate is available (and that a singular Jacobian is avoided). If the accurate initial estimate is not available, continuation or time-stepping techniques may slacken the solution process as was the case with the explicit time integration.

Finally, the equation (2.3) may be solved directly by means of a non-linear relaxation method. If (2.3) sufficiently satisfies stability (monotonicity) conditions, simple relaxation methods may converge. Rather than a good global convergence rate, we may expect that local relaxations -such as point Gauss-Seidel methods- will be able to smooth the error. Multigrid in its nonlinear form FAS (the Full Approximation Scheme) [1] is a proper technique to accelerate the convergence of these relaxations. The nonlinear relaxation iteration with FAS seems the most direct way to solve (2.3) and it is this approach that we follow in this paper.
In a local relaxation sweep, for each Ω_{ij} one set of four equations,

$$(N_h\,(q_h))_{ij} = 0\,, \qquad (3.3)$$

is solved (approximately). For these four equations no natural ordering exists and -hence- we solve these equations simultaneously (collective relaxation). Now only the local linearization, i.e. linearizaion of (3.3) with respect to $(q_h)_{ij}$ is needed.

Besides the reduced sensitivity for accurate initial estimates [16], it is an additional advantage of nonlinear relaxation that the storage requirements are significantly less. Only 4*4-systems are solved and no global Jacobian matrix needs to be kept, whereas a global linearization requires 80*N real numbers to store the Jacobian; N is the number of cells in the mesh. Probably a global linearization only pays off in a final stage of the solution process for (2.3), when Newton's method guarantees quadratic convergence. For a further discussion of the choice between Newton-Multigrid or Multigrid-Newton see [5].

Independent of the type of linearization that is actually used, it is important to see the structure of the Jacobian matrix of (2.3). For the finite volume discretization, the global Jacobian matrix is assembled in a way that is similar to the assembling process for finite elements. The finite volume Jacobian is the sum of small block-2*2 cell wall matrices (for FEM: element matrices). The regular structure of the mesh used induces a block-5-diagonal structure in which each block entry itself is a 4*4 matrix. The Jacobian and the rhs are assembled simultaneously and for each cell wall, Γ_{ijk}, the entries $\pm f(q_{ij}, q_{ijk})$,

$$\pm A_{ijk}^{+} = \pm \frac{\partial}{\partial q_{ij}} f(q_{ij}, q_{ijk}) \text{ and } \pm A_{ijk}^{-} = \pm \frac{\partial}{\partial q_{ijk}} f(q_{ij}, q_{ijk})$$

are evaluated and added as a contribution to the rhs or the Jacobian matrix respectively (see figure 1). The fact that all column sums are zero, except for boundary contributions, reflects that the discretization is conservative.
A row from the Jacobian, corresponding to the cell Ω_{ij}, is now seen to be of the form

$$-A^{+}_{i-\frac{1}{2},j} \quad ,-A^{+}_{i,j-\frac{1}{2}} \quad ,\begin{matrix} -A^{-}_{i,j-\frac{1}{2}} + A^{+}_{i,j+\frac{1}{2}} \\ -A^{-}_{i-\frac{1}{2},j} + A^{+}_{i+\frac{1}{2},j} \end{matrix} \quad ,+A^{-}_{i,j+\frac{1}{2}} \quad ,+A^{-}_{i+\frac{1}{2},j} \tag{3.4}$$

This structure of the Jacobian matrix is to be exploited when relaxation methods for the system are analyzed.

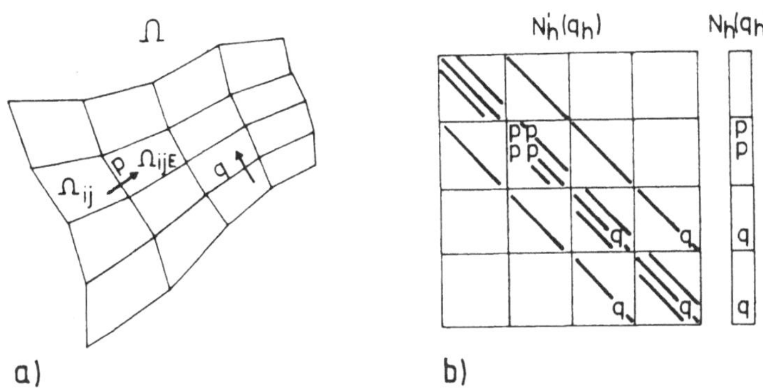

Figure 1. Assembling the rhs and the Jacobian of the nonlinear system.

In figure 1 the notation $\begin{bmatrix} P & P \\ P & P \end{bmatrix} \begin{bmatrix} P \\ P \end{bmatrix}$ is short for

$$\begin{bmatrix} +A^{+}_{ijE} & +A^{-}_{ijE} \\ -A^{+}_{ijE} & -A^{-}_{ijE} \end{bmatrix} \qquad \begin{bmatrix} +f(q_{ij},q_{ijE}) \\ -f(q_{ij},q_{ijE}) \end{bmatrix}, \tag{3.5}$$

where $A_{ijE} = A_{i,j+\frac{1}{2}}$.

The boundary condition treatment is straightforward as soon as we take the view that for each $\Gamma_{ijB} = \delta\Omega \cap \bar{\Omega}_{ij}$ we can determine a q_{ijB} as is introduced in section 2. This boundary state depends on the state in the neighbouring cell: $q_{ijB} = q_{ijB}(q_{ij})$. Hence, we have for the contribution from Γ_{ijB} to (2.3)

$$f_{ijB} := f(q_{ijB}(q_{ij}), q_{ij})$$

and

$$\begin{aligned} \frac{\partial}{\partial q_{ij}} f_{ijB} &= \frac{\partial}{\partial q_{ij}} f(q_{ijB}, q_{ij}) + \frac{\partial}{\partial q_{ijB}} f(q_{ijB}, q_{ij}) \ \frac{\partial}{\partial q_{ij}} q_{ijB}(q_{ij}) \\ &= A^{-}_{ijB} + A^{+}_{ijB} \cdot \frac{\partial q_{ijB}}{\partial q_{ij}} . \end{aligned} \tag{3.6}$$

This completes the description of the linearization of (2.3).

The local linearization of (3.3) gives a coefficient matrix as appears in the main diagonal of (3.4). For cells near $\delta\Omega$ these systems are augmented by boundary terms as given in (3.6).

4. THE NESTED SEQUENCE OF DISCRETIZATIONS

For the P-variant of Osher's scheme [3], when the flow field is sufficiently smooth (no shocks present), for any q_h defined on the grid $\Omega_h = \{\Omega_{ij}\}$, piecewise constant states q^*_{ijk} can be defined at the cell boundaries Γ_{ijk}, such that $f(q^*_{ijk}) = f(q_{ij}, q_{ijk})$ or, at the boundary of Ω, $q^*_{ijk} = q_{ijB}$. Under these assumptions the finite volume method can be seen as a formal weighted residual method for the discretization of $N(q)=0$.

We can write the discrete operator N_h as a Galerkin approximation to N,

$$N_h(q_h) = R_h N(P_h q_h), \tag{4.1}$$

where $P_h : X_h \to X$ relates to each q_h a function $P_h q_h$ on Ω, for which

$$\begin{cases} P_h q_n(s) = q_{ij} & \text{for } s \in \Omega_{ij} \\ P_h q_h(s) = q^*_{ijk} & \text{for } s \in \Gamma_{ij} \end{cases},$$

where q^*_{ijk} is such that $f(q^*_{ijk}) = f(q_{ij}, q_{ijk})$. I.e. piecewise constant functions with upwind continuity at the boundaries for characteristic information.
The restriction $\bar{R}_h : Y \to Y_h$ is defined by

$$(\bar{R}_h r)_{ij} = \int_{\Omega_{ij}} r(x,y)\,dxdy\ ,$$

so that, formally,

$$\begin{aligned}
(\bar{R}_h N(q))_{ij} &= \int_{\Omega_{ij}} f_x(q) + g_y(q)\,dxdy \\
&= \oint_{\delta\Omega_{ij}} n_1 f(q) + n_2 g(q)\,ds \\
&= \sum_k \int_{\Gamma_{ijk}} (n_1 f + n_2 g) q\ ds
\end{aligned}$$

and

$$\begin{aligned}
(\bar{R}_h N(P_h q_h))_{ij} &= \sum_k \int_{\Gamma_{ijk}} (n_1 f + n_2 q)(P_h q_h)\,ds \\
&= \sum_k \int_{\Gamma_{ijk}} (n_1 f + n_2 q)(q^*_{ijk})\,ds \\
&= \sum_k \text{meas}\,(\Gamma_{ijk}) f(q_{ij}, q_{ijk})\,.
\end{aligned} \tag{4.2}$$

Now, a regular sequence of nested discretizations is found in the following way. Start with a partitioning of Ω in cells $\{\Omega_{ij} | 0 < i \leq n_1 2^l, 0 < j \leq n_2 \cdot 2^l\}$. This yields the finest level of discretization (level l). For each $k = l, l-1, l-2, \ldots, 1$, a coarser level of discretization $k-1$ is defined by deleting each second meshline, so that each time 4 cells in the finer mesh correspond to a single coarser cell. In this way sequences of discrete spaces
$X_h, X_{2h}, X_{4h}, \cdots$, and $Y_h, Y_{2h}, Y_{4h}, \cdots$, are obtained. With a prolongation between the discrete solution spaces $P_{h,2h} : X_{2h} \to X_h$, defined by "piecewise constant interpolation" (distribute a coarse cell value q_{ij} as the same value over 4 finer cells), and a restriction $\bar{R}_{2h,h} : Y_h \to Y_{2h}$, defined by adding the 4 fine cell values to obtain the coarse cell value, we have the relation

$$P_{2h} = P_h P_{h,2h} \text{ and } \bar{R}_{2h} = \bar{R}_{2h,h} \bar{R}_h\ .$$

In this way we obtain a commutative diagram for the -now nested- set of discretizations N_h, N_{2h}, etc..

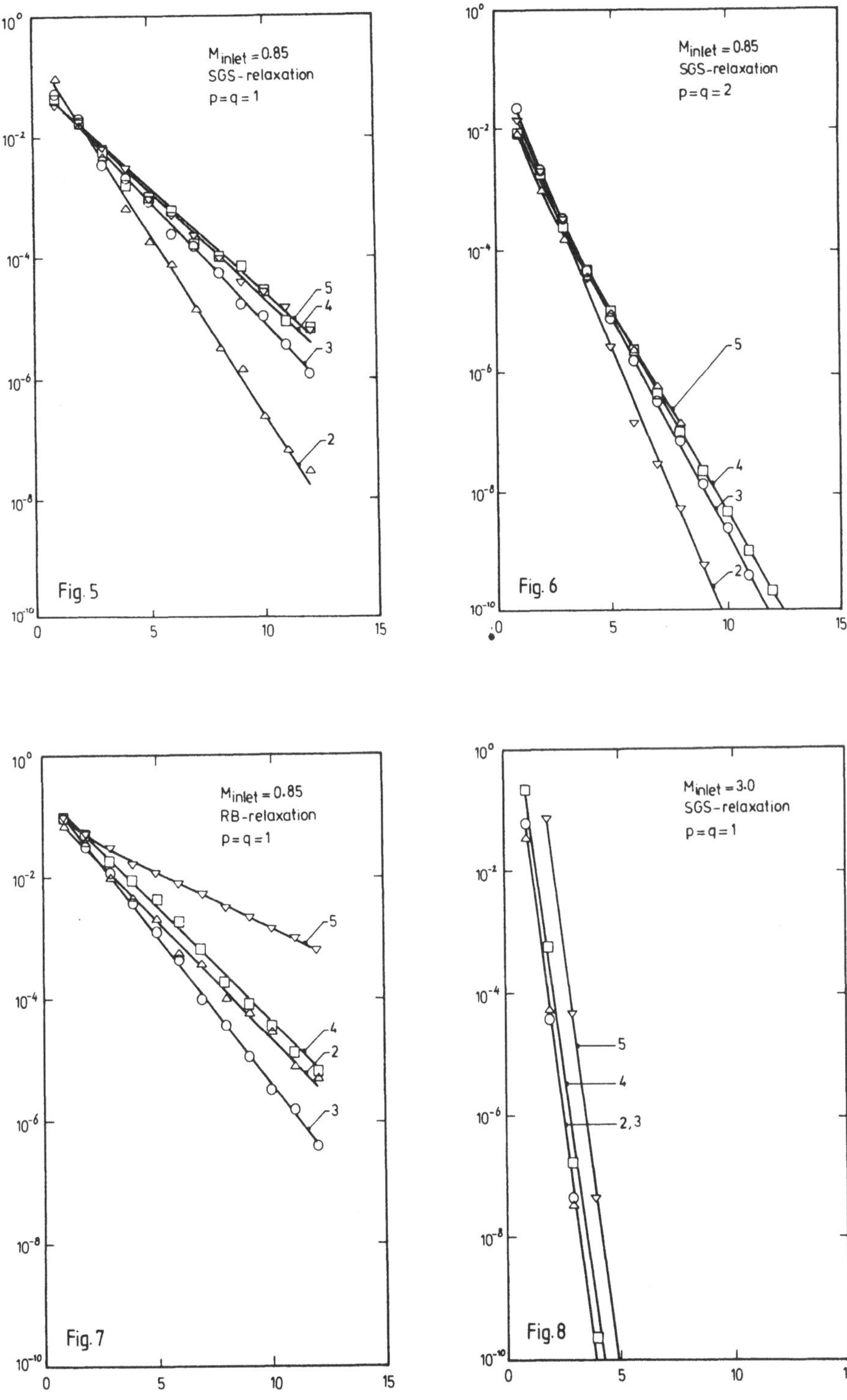

Figs. 5-8 Residual (ordinate) versus number of FAS-cycles (abscissa)

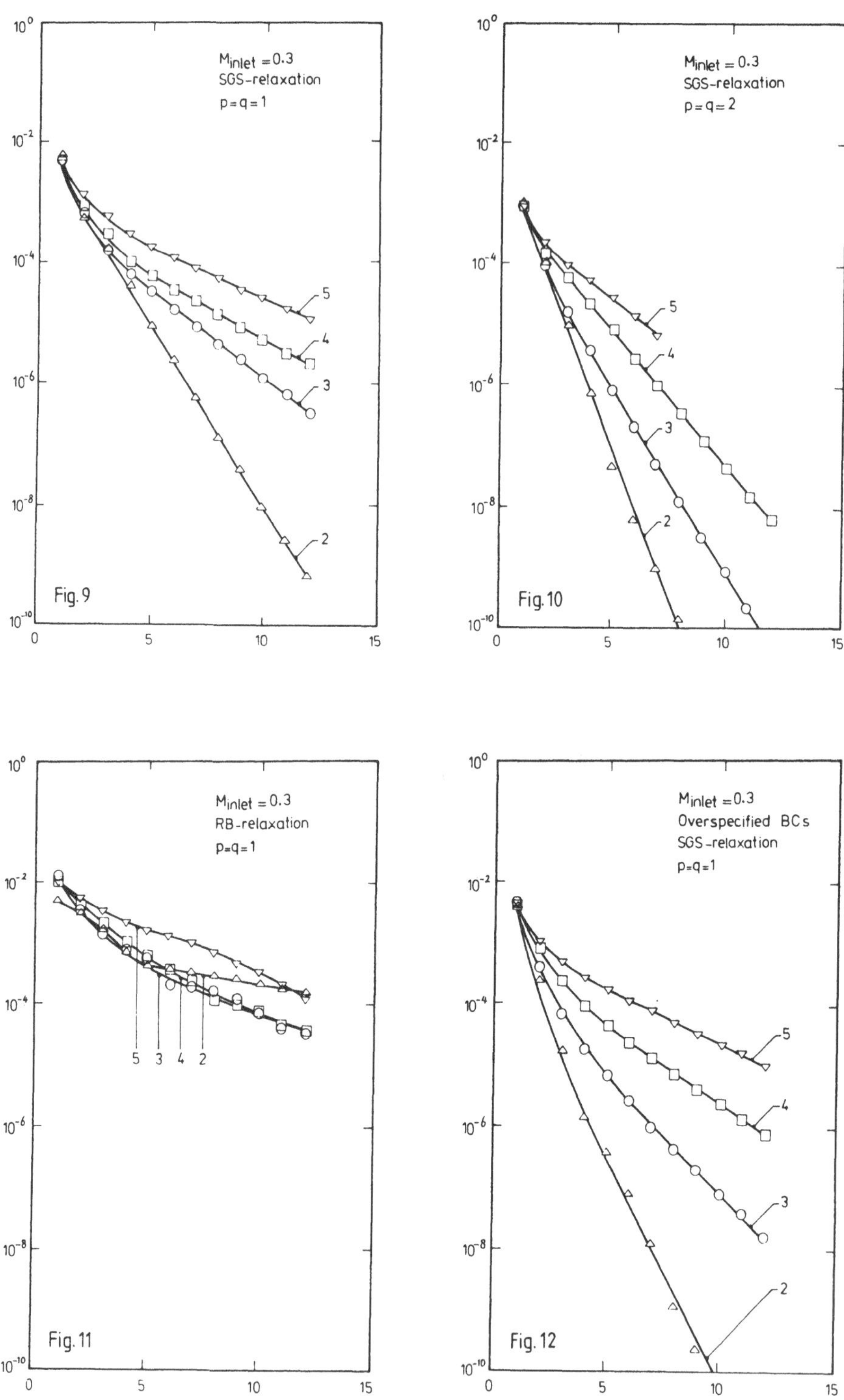

Figs. 9-12 Residual (ordinate) versus number of FAS-cycles (abscissa)

ANALYSIS OF A SOR-LIKE MULTI-GRID ALGORITHM FOR EIGENVALUE PROBLEMS

G. Hofmann

Institut für Informatik und Praktische Mathematik der Christian-Albrechts-Universität Kiel, Olshausenstr. 40, D-2300 Kiel 1, Germany

SUMMARY

In the recent years some multi-grid algorithms for the solution of eigenvalue problems have been developed. In this contribution, a SOR-like multi-grid method is proposed and the influence of perturbations in some algorithmic components is analysed. Finally, some practical experiences with this method are referred.

0. INTRODUCTION

There are some ways known how to apply multi-grid methods to eigenvalue problems of elliptic differential operators. In the last years three different approaches have been tried:

- inverse iteration with multi-grid methods (cf. Bank[1])
- nonlinear multi-grid methods (cf. Brandt, McCormick, Ruge[2])
- a SOR-like multi-grid method (cf. Hackbusch[3]).

The last method requires the solution of singular coarse-grid equations. The existence and uniqueness of these solutions is guaranteed by application of certain projections which are in effort of operations almost comparable to a smoothing step. A second disadvantage seems to be that the corresponding eigenvalues on the coarse grids have to be known exactly. The task of this work is to show that dropping the projections and some errors in the eigenvalues on coarse grids do not disturb the convergence of this method too much. In the first chapter the method will be described, in the second chapter the smoothing step will be analysed. For sake of simplicity, we restrict ourselves to one rather simple smoothing operator, the damped Jacobi iteration. Chapter 3 contains the analysis of the two-grid correction step. In chapter 4, the results of the multigrid analysis are given while chapter 5 contains some concluding remarks and applications of this analysis.

I. THE MULTI-GRID METHOD

As the underlying problem, we consider a linear symmetric eigenvalue problem: Let $\Omega \subset R^2$ be a C^∞-region.
We look for solutions u and λ of the following equation:

$$(L-\lambda B)u = 0 \text{ on } \Omega \tag{1}$$
$$+ \text{ boundary conditions for } u.$$

L is a differential operator of order 2 and B is an operator $B : L^2(\Omega) \to L^2(\Omega)$.

The discretisation of problem (1) leads to a set of problems

$$(L_h - \lambda_h B_h)u_h = 0 \text{ for } h \in]0,h_0]. \tag{2}$$

The boundary conditions of (1) are contained in the operators L_h and B_h.
We assume that all matrices L_h and B_h are symmetric.
Let $<.,.>$ denote the $L^2(\Omega)$ scalar product and $<.,.>_h$ the corresponding scalar product on the grid with gridsize h. Let $|u| := <u,u>^{1/2}$ and $|u_h|_h := <u_h,u_h>_h^{1/2}$ for all suitable functions u and u_h. To simplify the analysis, let us further assume that we have simple eigenvalues λ and λ_h with corresponding eigenfunctions e and e_h.
Let $<e,Be> = <e_h,B_he_h>_h = 1$.

The SOR-Method for solving equations of type (2) runs as follows:
Given some initial guess u_h^i and λ_h^i, we apply the following:

i) Compute u_h^{i+1} by application of some steps of an SOR-Method for solving the equation

$$(L_h - \lambda_h^i B_h)u_h = 0$$

with fixed λ_h^i.

ii) Compute λ_h^{i+1} from u_h^{i+1} by the Rayleigh quotient.

Let $h' := 2h$.
To get a multi-grid method, we replace step i) by one step of the corresponding multi-grid iteration using the first part of the former iteration as a smoother and correcting by means of the coarse grid equation

$$(L_{h'} - \lambda_{h'}B_{h'})u_{h'} = f_{h'} \tag{3}$$

$$\text{with } f_{h'} \in (e_{h'})^\perp, \; u_{h'} \in (B_{h'}e_{h'})^\perp. \tag{4}$$

We introduce the projections

$$Q_h^r := I - B_he_h<.,e_h>_h$$

and

$$Q_h^l := I - e_h\langle\cdot, B_h e_h\rangle_h \ .$$

Let $S_h(u_h,\lambda_h,f_h)$ be a smoothing operator depending on λ_h and f_h. r respectively p denote properly chosen restrictions and prolongations.

Then one step of the two-grid algorithm similar to the SOR-algorithm can be written as

$$u_h^{i+1/2} := S_h^{\nu}(u_h^i,\lambda_h,f_h);$$

$$f_h := (L_h - \lambda_h B_h)u_h^{i+1/2};$$

solve equation

$$(L_{h'} - \lambda_{h'}B_{h'})u_{h'} = Q_{h'}^r r f_{h'} \tag{5}$$

exactly with

$$u_{h'} \in (B_{h'}e_{h'})^{\perp} \ .$$

$$u_h^{i+1} := u_h^{i+1/2} - p u_{h'} \ .$$

In the usual manner, we replace the solving of equation (5) by a multi-grid algorithm.

Let step-sizes $h_0 < h_1 < h_2 < \ldots < h_l$ and the eigenvalues λ_h and eigenfunctions e_h for all $h \in \{h_0,h_1,\ldots,h_l\}$ be given. We write λ_i instead of $\lambda_{h_i}, \ldots$.

Then we use the following ALGOL-like recursive algorithm to solve (5):

```
PROCEDURE mgm (i,u,f); VALUE i; INTEGER i; ARRAY u,f;
IF i=0 THEN solve (5) by some other method
       ELSE
       BEGIN INTEGER j; REAL h,h'; ARRAY w,g;
       h := h_i; h':=h_{i-1}; w:=0.0;
       u := S_h^ν(u,λ_h,f);
       g := Q_h^r r((L_h - λ_h B_h)u - f);
       FOR j := 1 STEP 1 UNTIL γ DO mgm(i-1,w,g);
       u := Q_h^l(u - pw)
       END.
```

ν and γ are properly chosen integers.

A proof of convergence for this algorithm can be found in Hackbusch [3].

The eigenvalues λ_h and eigenfunctions e_h for $h \in \{h_0, \ldots, h_{l-1}\}$ and the starting values on grid h are computed by a "nested iteration":

Compute λ_0, e_0 by some other method; (6)

FOR i := 1 STEP 1 TO l DO
BEGIN

$u_i^0 := pe_{i-1};\ \lambda_i^0 := \lambda_{i-1};$

REPEAT one step of the multi-grid algorithm m TIMES;

$u_i := u_i^m;\ \lambda_i := \lambda_i^m.$

END.

m can be chosen independently of h to get a certain accuracy of λ_h and e_h on each grid.

In practical applications of this algorithm, we want to drop the projections Q_h^r and Q_h^l. This leads to a disturbed coarse-grid equation (5). The iterates u_h^i contain therefore some coefficient in direction of the eigenvector e_h.

To handle these perturbations, we split f_h and u_h in the following way

$$f_h := \tilde{f}_h + \alpha_h e_h \text{ with } \tilde{f}_h \in (e_h)^\perp \qquad (7)$$

and

$$u_h^i := u_h + v_h^i + \alpha^i e_h \text{ with } v_h^i \in (e_h)^\perp \text{ and } \alpha^i := \langle u_h^i, B_h e_h \rangle. \qquad (8)$$

Further, we assume that we do not know the exact eigenvalue λ_h but only an approximation $\tilde{\lambda}_h$.

We denote the error of this approximation by

$$X := |\tilde{\lambda}_h - \lambda_h|.$$

In the next chapters, we shall determine how the values $|v_h^{i+1}|_h$ and α^{i+1} depend on $|v_h^i|_h$ and α^i. The computations will show that we also need the values $\alpha_h, |u_h|_h$, and X to get a complete estimation of the desired values. To write these estimations in a simple form, we determine values $a_{i,j} \in \mathbb{R}_+$, i=1,2, j=1,...,4 such that the follwing is true:

$$\begin{bmatrix} |\alpha^{i+1}| \\ |v_h^{i+1}|_h \\ |\alpha_h| \\ |u_h|_h \end{bmatrix} \leq \begin{bmatrix} a_{11}(X) & a_{12}(X) & a_{13}(X) & a_{14}(X) \\ a_{21}(X) & a_{22}(X) & a_{23}(X) & a_{24}(X) \\ 0 & 0 & 1 & 0 \\ 0 & 0 & 0 & 1 \end{bmatrix} \begin{bmatrix} |\alpha^{i}| \\ |v_h^{i}|_h \\ |\alpha_h| \\ |u_h|_h \end{bmatrix} \tag{9}$$

The last two rows of this matrix show that $|u_h|_h$ and α_h remain unchanged during the iteration.

II. ANALYSIS OF THE SMOOTHING STEP

To simplify the calculations, we use a rather simple-structured smoother, the damped Jacobi iteration:

$$S(u_h, \tilde{\lambda}, f_h) = (I - \omega_h h^2 (L_h - \tilde{\lambda}_h B_h)) u_h^i - \omega_h h^2 f_h$$

with

$$\omega_h \leq 1/(h^2 |L_h - \tilde{\lambda}_h B_h|_{h \leftarrow h}).$$

Using (7) and the definitions of chapter I we obtain

$$\tilde{u}_h := (I - \omega_h h^2 (L_h - \lambda_h B_h))(u_h + v_h^i + \alpha^i e_h) + \omega_h h^2 f_h + \omega_h h^2 \alpha_h e_h$$

$$- \omega_h h^2 (\tilde{\lambda}_h - \lambda_h) B_h (u_h + v_h^i + \alpha^j e_h).$$

If we split this sum in a proper manner, we get the following terms

1) $(I - \omega_h h^2 (L_h - \lambda_h B_h)) u_h + \omega_h h^2 f_h = u_h.$

2) $|(I - \omega_h h^2 (L_h - \lambda_h B_h)) v_h^i|_h \leq g |v_h^i|_h$ with $g < 1.$

   where g denotes the convergence rate of the smoothing procedure,

3) $(I - \omega_h h^2 (L_h - \lambda_h B_h)) \alpha^i e_h = \alpha^i e_h.$

4) $\omega_h h^2 \alpha_h e_h$ remains unchanged.

5) $|\omega_h h^2(\tilde{\lambda}_h-\lambda_h)B_h(u_h + v_h^i + \alpha^i e_h)|_h \leq c\omega_h h^2 X|u_h + v_h^i + \alpha^i e_h|$.

Splitting $\tilde{u}_h^i = u_h + \tilde{v}_h^i + \tilde{\alpha}^i e_h$ as in (8), we get the estimate

$$\begin{bmatrix} |\tilde{\alpha}^j| \\ |\tilde{v}_h^j|_h \\ |\alpha_h| \\ |u_h| \end{bmatrix} \leq \begin{bmatrix} 1+c\omega_h h^2 X & c\omega_h h^2 X & \omega_h h^2 & c\omega_h h^2 X \\ c\omega_h h^2 X & g+c\omega_h h^2 X & 0 & c\omega_h h^2 X \\ 0 & 0 & 1 & 0 \\ 0 & 0 & 0 & 1 \end{bmatrix} \begin{bmatrix} |\alpha^i| \\ |v_h^i| \\ |\alpha_h| \\ |u_h| \end{bmatrix} \qquad (10)$$

If we use ν steps of this smoothing procedure, we have to raise this matrix to the power ν. This gives

$$\begin{bmatrix} 1 & 0 & \nu\omega_h h^2 & 0 \\ 0 & g^\nu & 0 & 0 \\ 0 & 0 & 1 & 0 \\ 0 & 0 & 0 & 1 \end{bmatrix}$$

$$+cXh^2 \begin{bmatrix} \nu & 2\nu & 0 & \nu \\ 2\nu & \nu g^{\nu-1} & 0 & G_\nu \\ 0 & 0 & 0 & 0 \\ 0 & 0 & 0 & 0 \end{bmatrix} \qquad (11)$$

$$+ O(\nu^3 h^6 X^2) + O(\nu^2 h^2 X^2) + O(h^4) + O(\nu^2 h^4 X) \text{ with } G_\nu := \sum_{i=0}^{\nu-1} g^i.$$

III. ANALYSIS OF THE TWO-GRID STEP

Let $h':=2h$ and $j:=i+1/2$. For all $\lambda\in\mathbb{R}$ and $h\in\{h_o,\ldots,h_1\}$ we define

$$L_h(\lambda) := L_h-\lambda B_h.$$

Let

$$K_{h',h}(\lambda_{h'}) := p\Omega^{1}_{h'}L^{-1}_{h'}(\lambda_{h'})\Omega^{r}_{h'}r.$$

If we use notation (7) for f_h and (8) for u^j_h, we can write the result of one two-grid correction step as follows:

$$u^{i+1}_h = u^j_h - K_{h',h}(\lambda_{h'})(L_h(\tilde{\lambda}_h)u^j_h - f_h) \tag{12}$$

$$= u_h+ v^j_h+ \alpha^j e_h-K_{h',h}(\lambda_{h'})(L_h(\tilde{\lambda}_h)(u_h+ v^j_h+ \alpha^j e_h)-\tilde{f}_h-\alpha_h e_h)$$

$$= \{u_h-K_{h',h}(\lambda_{h'})(L_h(\lambda_h)u_h-\tilde{f}_h)\}-K_{h',h}(\lambda_{h'})(\tilde{\lambda}_h-\lambda_h)B_h u_h+v^j_h$$

$$-K_{h',h}(\lambda_{h'})L_h(\lambda_h)v^j_h-K_{h',h}(\lambda_{h'})(\tilde{\lambda}_h-\lambda_h)B_h v^j_h+\alpha^j e_h$$

$$-K_{h',h}(\lambda_{h'})L_h(\lambda_h)\alpha^j e_h-K_{h',h}(\lambda_{h'})(\tilde{\lambda}_h-\lambda_h)B_h\alpha^j e_h$$

$$-K_{h',h}(\lambda_{h'})\alpha_h e_h.$$

Under the usual assumption for multi-grid convergence, the following is true: Let z_h be an arbitrary grid function on the grid with size h. Then for $z'_h := K_{h',h}(\lambda_{h'})z_h$, we have

$|z'_h|_h\leq C(h/h')|z_h|_h$ and if we have the splitting $z'_h=z''_h+xe_h$ with $z''_h\in(e_h)^{\perp}$, then $|z''_h|_h<C|z_h|_h$ and $|x|<Ch^2|z_h|_h$.

Using this we get the following estimates:

1) $u_h - K_{h',h}(\lambda_{h'})(L_h(\lambda_h)u_h - \tilde{f}_h) = u_h$.

2) $|K_{h',h}(\lambda_{h'})(\tilde{\lambda}_h-\lambda_h)B_h u_h|_h\leq CX|u_h|_h$.

3) $|v^j_h - K_{h',h}(\lambda_{h'})L_h(\lambda_h)v^j_h|_h\leq K|v^j_h|_h$.

The constant K is induced by the "approximation property" of the discretisation.

4) $|K_{h',h}(\lambda_{h'})(\tilde{\lambda}_h-\lambda_h)B_h v_h^j|_h \leq CX|v_h^j|$.

5) $K_{h',h}(\lambda_{h'})L_h(\lambda_h)\alpha^j e_h=0$.

6) $|K_{h',h}(\lambda_{h'})(\tilde{\lambda}_h-\lambda_h)B_h\alpha^j e_h|_h \leq CX\alpha^j$.

7) $\alpha^j e_h$ remains unchanged.

8) $|K_{h',h}(\lambda_{h'})\alpha_h e_h|_h \leq C\alpha_h$.

Using the same method as in chapter II, we can collect these estimates in a matrix.

$$\begin{bmatrix} 1 & 0 & Ch^2 & 0 \\ 0 & K & C & 0 \\ 0 & 0 & 1 & 0 \\ 0 & 0 & 0 & 1 \end{bmatrix} + CX \begin{bmatrix} Ch^2 & Ch^2 & 0 & Ch^2 \\ 1 & 1 & 0 & 1 \\ 0 & 0 & 0 & 0 \\ 0 & 0 & 0 & 0 \end{bmatrix} \tag{13}$$

To get the complete error-matrix for one smoothing-correction step,we have to multiply matrices (9) and (13). Using the abbreviation $p_h := \nu\omega_h h^2$ we get

$$\begin{bmatrix} 1 & 0 & p_h & 0 \\ 0 & Kg^\nu & \omega_h h^2 KG_\nu & 0 \\ 0 & 0 & 1 & 0 \\ 0 & 0 & 0 & 1 \end{bmatrix} +$$

$$CX \begin{bmatrix} p_h+Ch^2 & 2p_h+Ch^2g^\nu & 2p_h(\nu-1)h^2 & p_h(1+(\nu-1)h^2 \\ 2p_hK+1 & p_hKg^{\nu-1}+g^\nu & p_h+h^2(K\nu(\nu-1)+G_\nu)) & \omega_h h^2 KG_\nu \\ 0 & 0 & 0 & 0 \\ 0 & 0 & 0 & 0 \end{bmatrix}$$

Generally, the two-grid convergence rate is different from the product Kg^{ν} which we get if we use this kind of analysis. To take the correct rate into account, we have to replace the product by this rate. To finish the analysis of the two grid step, we again have to raise the error matrix to the power γ. This leads to

$$\begin{bmatrix} 1 & 0 & p_h & 0 \\ 0 & k^{\gamma} & Ch^2 & 0 \\ 0 & 0 & 1 & 0 \\ 0 & 0 & 0 & 1 \end{bmatrix} + CX \begin{bmatrix} h^2 & h^2 & h^2 & h^2 \\ 1+h^2 & 1+h^2 & h^2 & 1 \\ 0 & 0 & 0 & 0 \\ 0 & 0 & 0 & 0 \end{bmatrix} \qquad (14)$$

$$+ O(X^2).$$

In the second matrix, p_h is replaced by $O(h^2)$ under the assumption that ν is uniformly bounded with respect to h.

IV. RESULTS OF THE MULTI-GRID ANALYSIS

As in matrix (14), we restrict ourselves in the multi-grid analysis to linear terms of the eigenvalue errors. If we make this simplifying assumption, we can assemble the error matrix in the following way.

Let $X_k := |\tilde{\lambda}_k - \lambda_k|$ for $0<k\leq i$ and Z_i be the error matrix on level i.

Then Z_i can be estimated by

$$Z_i \leq C(A_i + (\gamma X_i)/2*B_i + (\gamma^2 X_{i-1})/4*C_i + \sum_{k=0}^{i-2} (\gamma/2)^{i-k+1} X_k D_{i,k}) + O(X_r X_s)$$

for $r,s\in\{1,\ldots,i\}$, where $A_i, B_i, C_i, D_{i,k}$ denote the error matrices for the following cases:

A_i : on all levels exact eigenvalues are given,
B_i : only on the finest level i we have an eigenvalue error,
C_i : only on level i-1 we have an eigenvalue error,
$D_{i,k}$: finest level is i and on level k, $0<k<i-1$, we have an eigenvalue error.

We only want to give a short outline to the method of the estimation. Using the first part of matrix (14) as starting value we can construct a recursive formula for A_i by induction.

Taking into account the eigenvalue error on level i we can compute matrix B_i using error matrix A_{i-1} for the coarse-grid estimation. The same way, using B_{i-1} for coarse-grid estimation we can compute C_i. As in the computation of A_i, taking C_{k+1} as starting value we get the matrices $D_{i,k}$. If we make some more simplifying assumption, we can resolve the recursive formulas and get explicit bounds for the errors. Because these bounds are of some practical interest, we shall list them here. For the recursive formulas and proofs, see Hofmann, G. [5].

$$A_i \leq \begin{bmatrix} 1 & i\gamma^{i+1}p_0 & \gamma^i p_0 & 0 \\ 0 & 1 & 1 & 0 \\ 0 & 0 & 1 & 0 \\ 0 & 0 & 0 & 1 \end{bmatrix}$$

$$B_i \leq \begin{bmatrix} i\gamma^{i+1}p_0 & i\gamma^{2i+1}p_0^2 & i\gamma^{2i}p_0^2 & i\gamma^{i+1}p_0 \\ 1 & i\gamma^{i+1}p_0 & \gamma^i p_0 & 1 \\ 0 & 0 & 0 & 0 \\ 0 & 0 & 0 & 0 \end{bmatrix}$$

$$C_i \leq \begin{bmatrix} 0 & i^2\gamma^{2i+1}p_0^{\,2} & i^2\gamma^{3i-2}p_0^2 & 0 \\ 0 & (i-1)\gamma^i p_0 & i\gamma^i p_0 & 0 \\ 0 & 0 & 0 & 0 \\ 0 & 0 & 0 & 0 \end{bmatrix}$$

$$D_{i,k} \leq \begin{bmatrix} 0 & p_0^2(k^2\gamma^{3k} + \sum_{j=k+1}^{i} ij\gamma^{i+j}) & p_0^2(k^2\gamma^{3k} + \sum_{i=k+1}^{i} ij\gamma^{i+j}) & 0 \\ 0 & 2^{i-k}k\gamma^k p_0 & 2^{i-k}k\gamma^k p_0 & 0 \\ 0 & 0 & 0 & 0 \\ 0 & 0 & 0 & 0 \end{bmatrix}$$

V. CONCLUDING REMARKS

V.1 Choice of the smoothing operator

The estimates given in the last chapter allow two obvious conclusions:

- The number of coarse grid iterations γ should not be too large (e.g. $\gamma=1,2$) which also is reasonable from the aspect of efficiency.
- If the number of levels i is not too large and h_0 sufficiently small, then convergence of the disturbed method is induced by convergence of the undisturbed method.

A crucial point in development of multi-grid methods is the choice of a good smoothing operator. During the analysis we restricted ourself to the rather simply structured case of damped Jacobi iteration, which allowed a strict estimation of the occuring errors. The results give also some ideas how to choose smoothing operators. If we have an error matrix for the smoothing step like (9), comparison with (10) shows that $a_{11}(X)$ and $a_{22}(X)$ can be written as

$a_{11}(X)=1+a'_{11}(X)$; $a_{22}(X)=g+a'_{22}(X)$ with $a'_{11}(X) \to 0$ and $a'_{22}(X) \to 0$ as $X \to 0$. This represents the fact that the eigenfunction is a fixed point of the smoothing operator if we have the exact eigenvalue and that the undisturbed smoothing operator diminishes or at least smoothes the error v_h^i (compare (8)). The calculation of the powers of matrix (10) show that $a_{12}(X)$ and $a_{21}(X)$ should be small if X is small. This is an advantage of the damped Jacobi iteration since $a_{12}(X) = a_{21}(X) = 0$ for all $X \in \mathbb{R}$ in the important case that $B_h = I$. This is caused by the fact that the eigenfunctions of L_h and of the smoothing operator are equal. If $B_h \neq I$, the functions $a_{12}(X)$ and $a_{21}(X)$ are bounded by $\omega_h h^2 X |B_h|_{h<-h}$ where $|\ |_{h<-h}$ denotes the operator-norm of B_h according to the $|\ |_h$ - norm.

V.2 Calculating higher eigenvalues

It is possible to compute higher eigenvalues by choosing another starting value in the nested iteration (6). The coarse-grid equation (compare (5)) becomes then indefinit which can cause divergence of the method. There are some possibilities to deal with this situation:

- diminishing of the coarsest step size h_0,
- choosing of a smoothing operator which fits for indefinit problems for example the modified Jacobi iteration

$$u_h - (I - \omega_h^2 h^4 L_h^2(\lambda_h)) u_h + \omega_h^2 h^4 L_h(\lambda_h) f_h. \tag{15}$$

- Simultaneous computation of the first eigenvalues and eigenfunctions together with orthogonalisation and Ritz projection.

Practical results show that in most cases the first method if applicable (enough storage available) gives the best results. It can also be shown by very simple examples that the Gauss-Seidel iteration does not fit as a smoothing operator for the computation of higher eigenvalues whereas the Jacobi iteration shows good results. The modified Jacobi iteration brings slow convergence but is rather robust (compare Hackbusch, Hofmann [4]).

V.3 Computation of clustered eigenvalues

If some eigenvalues are situated very near to each other, two problems appear

1) The choice of the starting value for the eigenvalue on the new grid in the nested iteration.
2) The solution of the very bad conditioned linear system (5) on the coarsest grid.

To deal with the first problem there are some possibilities:

- computation of the starting value by extrapolation
- using the prolongated eigenvector to compute the starting value by the Rayleigh quotient.

The difficulty of solving the linear system can be overcome by using a proper iteration method to solve (5). In most cases, about 4-6 iterations of the modified Jacobi iteration (15) gave good results.

V.4 Nonsymmetric eigenvalue problems

In the case of a nonsymmetric eigenvalue problem a similar algorithm can be constructed using left and right eigenfunctions. The analysis of chapters 2-4 can also be done for this case but is somewhat more complicated. The effort of computing time and storage to run the iteration is doubled. But there is also the possibility to use the algorithm of Chapter 1. The worse errors for the Rayleigh quotient lead to a slower convergence. Because this algorithm needs only half of the storage, it might be useful in some cases to use the simpler version.

V.5 Final remarks

The results of the error estimates in chapter 2-4 can be shown to be sharp by very simple examples. To get explicit results for more complicated situations, say in the case of other smoothing procedures than the Jacobi iteration, it might be useful to calculate the error matrices on a computer. The analysis can, in a slightly different form, also be applied to multi-grid algorithms for solving singular equations which may arise in other applications, for example in continuation procedures. The test of smoothing procedures brought the result that the damped Jacobi iteration is a very robust smoother which works even in the case of higher eigenvalues. Other smoothing procedures, e.g., the Gauss-Seidel iteration, may be much more efficient in calculating the first eigenvalue. But usually they fail in computing higher eigenvalues. Some care should also be spent in solving the coarse-grid equation on the coarsest grid, especially in the case of closely situated eigenvalues. Then also the nested iteration should be modified to get better starting values on the finer grids. The error analysis showed that the projections can be dropped without loosing too much which also agrees with practical experiences. The iteration error of the eigenvalues which arises in the nested iteration procedure, has also no dramatical effect on the convergence. Some results of this method for the plate equation can be found in Hackbusch, Hofmann [4].

REFERENCES

[1] Bank, R.: Analysis of a multilevel inverse iteration procedure for eigenvalue problems. SIAM J. Numer. Anal., Vol. 19, 886-893 (1982).

[2] Brandt, A., McCormick, S., Ruge, J.: Multigrid methods for differential eigenproblems. SIAM, J. Sci. Statist. Comput. 4, 244-260 (1983).

[3] Hackbusch, W.: On the computation of approximate eigenvalues and eigenfunctions of elliptic operators by means of a multi-grid method. SIAM J. Numer. Anal., Vol. 16, 201-215 (1979).

[4] Hackbusch, W., Hofmann, G.: Results of the eigenvalueproblem for the plate equation. Zeitung für Angew. Math. Phys., Vol. 31, 730-739 (1980).

[5] Hofmann, G.: Analyse eines Mehrgitterverfahrens zur Eigenwertberechnung. Doctoral thesis, Inst. f. Informatik und Prakt. Mathematik, Christian-Albrechts-Universität Kiel, (1985).

A MULTIGRID TREATMENT OF STREAM FUNCTION NORMAL DERIVATIVE BOUNDARY CONDITIONS

H. Holstein and G. Papamanolis
Department of Computer Science, U.C.W. Aberystwyth
Dyfed, SY23 3BZ, U.K.

SUMMARY

Numerical treatments of two-dimensional viscous flow commonly discretise the stream function normal derivative boundary conditions in terms of a boundary vorticity. We investigate four methods of relaxing such a boundary condition, within the context of a multigrid treatment employing pointwise relaxation of the steady creeping flow within a driven capacity. For each method, we calculate the average error reduction factor per multigrid iteration, and in some cases are able to obtain ideal values characteristic of the interior problem.

INTRODUCTION AND STATEMENT OF THE PROBLEM

We consider the application of the multigrid method to the solution of creeping flow in a driven cavity of unit square cross-section (see Fig.1):

$$\Delta\psi-\omega = 0 \tag{1}$$

$$\Delta\omega = 0, \text{ on interior of unit square,} \tag{2}$$

$$\psi = 0 \tag{3}$$

$$\partial\psi/\partial n = f, \text{ on boundary of unit square} \tag{4}$$

$$\text{where } f = \begin{cases} 1 \text{ on } CD, \\ 0 \text{ on } AB, BC, DA \end{cases} \tag{5}$$

and $\partial/\partial n$ represents the derivative along the inward normal $\underline{n}$.

We express (4) in discretised form using the method of Thom [1,2]

$$\frac{\partial\psi}{\partial n} = \frac{\psi(\underline{n}h)}{h} - \frac{h}{2}\omega + O(h^2) = f \tag{6}$$

at non-corner boundary points of the unit square. Here, h is the discretisation step length. In (6) and subsequent equations, all grid functions without arguments are evaluated at a boundary grid point. Where an argument is present, it refers to a position relative to the boundary point.

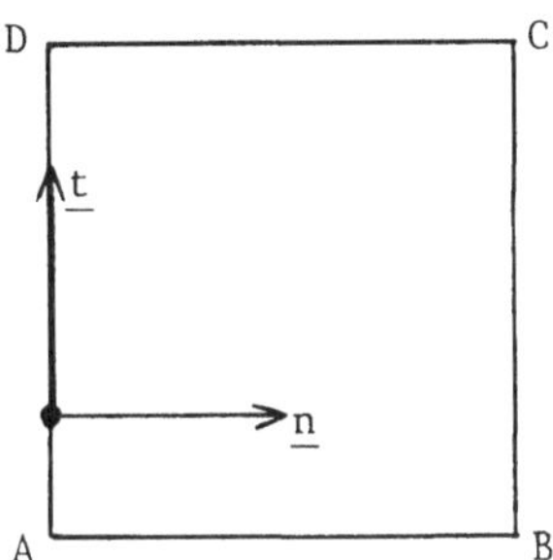

Fig.1 Square domain ABCD, with unit inward normal vector $\underline{n}$ and unit tangential vector $\underline{t}$ at a typical boundary point.

A multigrid algorithm for the system (1),(2) subject to boundary conditions (3),(6) has much in common with that used for a coupled Poisson system with Dirichlet boundary conditions. Thus we employed the following multigrid components:

(i) 5-point discretisation of the Laplacians for ψ and ω at interior points of a regular mesh
(ii) standard coarsening
(iii) fixed multigrid algorithm on 6 levels using V- or W-cycles
(iv) restriction operators corresponding to either injection or full weighting (INJ or FW)
(v) bilinear interpolation
(vi) pointwise collective Gauss-Seidel relaxation at interior points for ψ and ω.

A further component, namely

(vii) pointwise boundary relaxation of (6)

has no analogue in the Dirichlet case.

As pointed out by Brandt [3] one must (a) devise a suitable boundary relaxation scheme which smooths the error along the boundary while preserving the interior smoothness, or (b) use full residual weighting at the boundary, particularly also when (a) is not satisfied.

In this paper we follow approach (a). We propose four methods of boundary relaxation, and compare them by computing the average error reduction factor per multigrid iteration for each method. The results are summarised in Table 2.

Since the discretised system is linear, the reduction in error norms between consecutive multigrid iterations is most conveniently computed without loss of generality from the homogeneous problem, characterised by replacing (5) by

$$f = 0, \tag{7}$$

with interior solution $\psi=\omega=0$. The initial non-zero iterate for ψ and ω then forms the initial error, and subsequent iterates for ψ and ω then give the error directly. By averaging the error reduction over many iterations, we effectively estimate the norm-independent spectral radius of the multigrid iteration matrix, using a method similar to the inverse power method [4]. Implementation details are to be found in [5].

ESTIMATES FOR ACHIEVABLE ERROR REDUCTION FACTORS

We are guided by Brandt [3] to seek algorithms whose treatment of the boundary does not degrade the error reduction factors predicted by the local mode analysis for interior points far from the boundary. Such an analysis for the coupled equations (1),(2), using point-collective Gauss-Seidel and a 5-point Laplace discretisation, is easily shown to reduce to the corresponding local mode analysis for a single Poisson equation, extensively treated in [6,7]. We give the error reduction factors expected on such a basis in Table 1. We also include the corresponding 2-grid convergence factors [7] for a single Poisson system, and the computed average error reduction factors for the coupled system (1),(2) with Dirichlet boundary conditions for both ψ and ω. The expected agreement between the latter two columns is found, and close correlation with the local mode analysis is evident. We thus obtain a clear indication of the error reduction factors we wish to achieve when boundary relaxation of (6) is incorporated into the algorithm.

BOUNDARY RELAXATION METHODS

Below we give four methods of relaxing the boundary condition (6). A boundary relaxation sweep is carried out at all non-corner points taken in lexicographic order, that is, from A to B, from A to D and from B to C pointwise alternatively, and from D to C. We distinguish between Version 1 and Version 2 of each method, according to whether boundary relaxation is carried out before or after interior relaxation.

For the finest grid, f in equation (6) takes prescribed values (given by (7) in the homogeneous case). On coarse grids, f takes on the restriction of the residual values of (6) computed from the next finer grid. The residual will in fact be zero in Version 1 of Methods 3 and 4 and in Version 2 of Methods 1,3 and 4, and so residual transfer of (6) is not needed in these cases. When residual transfer is necessary, the scaling implied by (6) is used throughout.

Method 1

At all non-corner points, solve

$$\frac{\psi(\underline{n}h)}{h} - \frac{h}{2}\omega = f \tag{8}$$

for ω. Only the first boundary sweep yields distinct information.

Method 2

Define the boundary residual r to be

$$r = f - \left(\frac{\psi(\underline{n}h)}{h} - \frac{h}{2}\omega\right) \tag{9}$$

at all non-corner points, and to be zero at corner points. Then, for each non-corner point in turn, solve

$$r(\underline{t}h) - 2r + r(-\underline{t}h) = 0 \tag{10}$$

for ω. The motivation for such a scheme is given in [3].

Method 3

At all boundary points not adjacent to a corner, solve for ω, $\psi(\underline{n}h)$, $\omega(\underline{n}h)$, using a 3×3 matrix system defined by (8) and discretised forms of (1) and (2) centred at $\underline{n}h$.

Let unit vectors $\underline{n}'$ and $\underline{t}'$ be defined at a given corner, to lie along the boundaries. Then boundary values of ω at $\underline{n}'h$ and $\underline{t}'h$, and interior values of ψ and ω at $(\underline{n}'+\underline{t}')h$ may be solved from a 4×4 matrix system defined by (8) at $\underline{n}'h$, $\underline{t}'h$ and discretised forms of (1) and (2) centred at $(\underline{n}'+\underline{t}')h$.

Method 4

Consider a regular fine mesh of $(2^n+3) \times (2^n+3)$ points superimposed on the unit square domain. Standard coarsening can now be applied to the interior mesh of $(2^n+1) \times (2^n+1)$ points, in a way that maintains a constant distance between interior points adjacent to the boundary and the boundary itself. The coarse boundary points are taken to lie on the continuation of the coarse mesh lines on to the boundary.

This method of grid coarsening is applied in Method 4, together with the boundary relaxation described for Method 3. The 3×3 and 4×4 systems, however, now employ progressively more non-uniform discretisations of the Laplacian at interior points adjacent to the boundary, as the coarsening proceeds.

Table 1 Error reduction factors for Dirichlet Poisson problems

ν_1,ν_2 - relaxations before and after coarse grid correction

ρ^* - 2-grid convergence factor [7]

h_f,h_c - finest and coarsest mesh sizes

Order of relax. Restriction	Cycle (ν_1,ν_2)	Single Poisson loc. mod.	Single Poisson ρ^*	eq. (1),(2) $h_f = 1/64$, $h_c = 1/2$
LEX,INJ	V(1,1)	0.25	0.20	0.19
	W(1,1)	0.25	0.20	0.20
LEX,FW	V(1,1)	0.25	0.19	0.18
	W(1,1)	0.25	0.19	0.19
	W(1,2),(2,1)	0.125	0.12	0.12
RB,FW	W(1,1)	0.0625	0.07	0.07

RESULTS AND CONCLUSIONS

It is clear from Table 2 that a suitable combination of full-weighting, lexicographic ordering, W-cycles and boundary relaxation methods 2-4 is able to produce error reduction factors predicted by the local mode analysis. Method 1 makes no attempt to smooth errors in ω along the boundary, and so its poor performance is not surprising.

Method 4 achieves the best error reduction factors, and is able to yield good results even when injection is used in place of full weighting. The good performance is consistent with Brandt's statement that residual weighting near boundaries is normally required. By ensuring that the interior points adjacent to the boundary retain a constant distance from the

Table 2 Computed error-reduction factors for boundary relaxation methods 1-4, averaged over 50 multigrid cycles, for the system (1),(2),(3),(6)

ν_1, ν_2, h_f, h_c: see legend for Table 1

V1,V2: Version 1 and Version 2 of a given method

-: entry not applicable to method

Order of relaxation Restriction	Cycle (ν_1,ν_2)	Number of boundary relaxation sweeps	$h_f = 1/64$, $h_c = 1/2$						$h_f = 1/66$, $h_c = 1/2-1/66$		Local mode
			Method 1		Method 2		Method 3		Method 4		
			V1	V2	V1	V2	V1	V2	V1	V2	
LEX INJ	V(1,1)	1	2.00	11.48	0.84	0.67	1.56	1.45	0.97	3.10	0.25
		6	-	-	1.41	1.58	0.76	1.39	0.94	3.43	
	W(1,1)	1	0.55	1.07	0.40	0.46	0.39	0.48	0.23	0.21	
		3	-	-	0.34	0.46	0.34	0.49	0.20	0.24	
		6	-	-	0.42	0.46	0.33	0.48	0.20	0.24	
LEX FW	V(1,1)	1	1.40	2.23	0.45	0.46	1.81	1.25	0.50	1.13	
		6	-	-	0.91	1.29	0.83	0.36	0.52	1.15	
	W(1,1)	1	0.43	0.43	0.34	0.35	0.36	0.33	0.20	0.20	
		2	-	-	0.30	0.32	0.27	0.18	0.20	0.20	
		3	-	-	0.28	0.29	0.26	0.19	0.20	0.19	
		6	-	-	0.25	0.25	0.26	0.19	0.20	0.19	
	W(2,1)	1	0.22	0.29	0.24	0.25	0.27	0.22	0.12	0.13	0.125
		2	-	-	0.21	0.22	0.18	0.11	0.12	0.13	
	W(1,2)	1	0.25	0.29	0.24	0.25	0.27	0.22	0.12	0.13	
		2	-	-	0.21	0.22	0.19	0.12	0.12	0.13	
RB FW	W(1,1)	1	0.48	0.47	0.26	0.28	0.37	0.34	0.13	0.13	0.0625
		2	-	-	0.23	0.24	0.30	0.21	0.10	0.15	

boundary during coarsening, the need for level dependent residual weighting is removed. This property is not shared by Methods 1-3.

Red-black ordering in conjunction with Methods 1-3 leads to no significant improvements. However, the overall best error reduction among the cases considered was obtained with Method 4 using red-black interior relaxation. The reduction factor is somewhat higher than would be expected from the local mode analysis, and some further improvement may be possible.

We note that the double specification of boundary condition (4) at corners leads to a loss of regularity for the problem. The poor performance of V-cycles with all methods may therefore indicate that the higher regularity assumptions made in V-cycle convergence proofs, compared to those required for W-cycles, are indeed necessary.

ACKNOWLEDGEMENTS

The authors gratefully acknowledge helpful discussions and correspondence with Professor A. Brandt and Dr J.Linden. The substance of the last paragraph was brought to the attention of the authors during discussion of this paper at the Oberwolfach conference.

REFERENCES

[1] THOM, A: "An investigation of fluid flow in two dimensions", Aer. Res. Ctee., R. and M., No.1194, United Kingdom, 1929.

[2] ROACHE, P.J.: "Computational Fluid Dynamics", Hermosa Publishers, 1972.

[3] BRANDT, A: "Guide to multigrid development", in Multigrid Methods, Lecture Notes in Mathematics 960, ed. Hackbusch, W. and Trottenberg, U., Springer Verlag, 1982.

[4] WILKINSON, J.H.: "The Algebraic Eigenvalue Problem", Clarendon Press, 1965.

[5] PAPAMANOLIS, G.: "Multigrid Methods in Fluid Dynamics", M.Sc. Thesis, University of Wales, 1984.

[6] BRANDT, A.: "Multilevel adaptive solutions to boundary-value problems", Mathematics of Computation, Vol.31, Number 138, April 1977.

[7] STÜBEN, K., TROTTENBERG, U.: "Multigrid methods", in Multigrid Methods, Lecture Notes in Mathematics 960, ed. Hackbusch, W. and Trottenberg, U., Springer-Verlag, 1982.

A MULTIGRID METHOD FOR SOLVING THE BIHARMONIC EQUATION ON RECTANGULAR DOMAINS

Johannes Linden

Universität-GHS-Essen, FB 10
Universitätsstraße, D-4300 Essen, West Germany

SUMMARY

A simple and direct application of multigrid techniques for solving a standard finite difference approximation for the first boundary value problem of the biharmonic equation is described. Numerical results show that the convergence factors of the iterative multigrid method are h-independent and correspond to the good interior smoothing factors of the relaxation used. Moreover the convergence factors of our method are comparable to those of corresponding multigrid iterations for the Poisson equation while its computational effort per iteration is higher by a factor of 2.3 - 3 (depending on the chosen variant). Furthermore numerical results for the full-multigrid method (FMG) are presented. As in the case of Poisson's equation, FMG turns out to be an efficient approximate direct solver for the biharmonic equation with complexity $O(N^2)$ when solving the problem on a NxN grid, yielding solutions up to discretization errors.

1. INTRODUCTION

We consider the first boundary value problem of the biharmonic equation

$$\begin{aligned} \Delta\Delta u &= f \quad (\Omega) \\ u &= g \quad (\partial\Omega) \\ u_n &= \overline{g} \quad (\partial\Omega) \end{aligned} \tag{1.1}$$

in a rectangular domain Ω; u_n denotes the outer normal derivative on $\partial\Omega$.

A widely used difference discretization of the continuous problem (1.1) is the 13-point-stencil

$$\frac{1}{h^4}\begin{bmatrix} & & 1 & & \\ & 2 & -8 & 2 & \\ 1 & -8 & 20 & -8 & 1 \\ & 2 & -8 & 2 & \\ & & 1 & & \end{bmatrix} \tag{1.2}$$

with quadratic or cubic extrapolation on the boundary |1|; here h denotes the meshsize of a square grid. The order of consistency of (1.2) is $O(h^2)$. Estimates of the global discretization error are given in |2|,|6|,|11|; they are summarized in |1|,|7|.

Several direct or iterative methods for solving this standard approximation of (1.1) are proposed in the literature. They have a theoretical complexity of at least $O(N^3)$ for direct solvers and of $O(N^{5/2}\log_2 N)$ for iterative solvers when using a NxN grid (e.g. |4|,|10|). The first "optimal" $O(N^2)$-method was developed by P. Bjorstad |1| in 1981.
Multigrid methods for the computation of difference approximations of (1.1) have not yet been published in the literature. In some of his papers (e.g. |3|), A. Brandt gives short hints for a suitable treatment of the problem in the multigrid context. He recommends solving it in the equivalent form of a system of two second order equations coupled by the boundary conditions:

$$\begin{aligned} \Delta v &= f \\ \Delta u - v &= 0 \end{aligned} \quad (\Omega) \qquad u = g \; ; \; u_n = \bar{g} \quad (\partial\Omega) \tag{1.3}$$

which makes it easier to find efficient smoothing procedures. Brandt also discusses briefly and in a more general context a special way of treating the boundary conditions |3|.
Recently numerical investigations were done by G. Papamanolis |8| with a number of multigrid methods for solving (1.3) using different boundary treatments. The results were presented in |5|. The simplest and probably the most efficient one of these methods is quite similar to our approach, which is described in the present paper.
We always start from the formulation (1.3) of the biharmonic problem. Its difference discretization (equivalent to (1.2)) is stated in section 2. The components of the multigrid method

for solving the resulting system of linear equations are given in section 3. In section 4 we summarize numerical results for the multigrid iteration (MGI) as well as for the full-multigrid method (FMG); operation counts are also contained in section 4.

2. THE DISCRETE PROBLEM

The domain $\Omega=[0,A]\times[0,B]$ is covered by a rectangular grid Ω_h with meshsizes $\Delta x=A/N$ and $\Delta y=B/M$ in x- and y-direction, $h=(\Delta x,\Delta y)$.

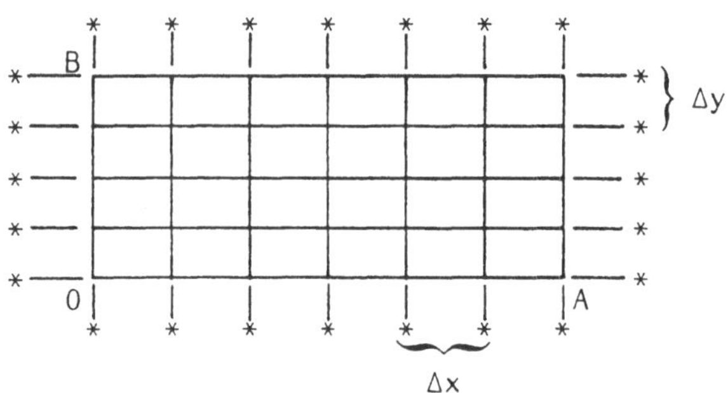

Fig. 2.1: The computational grid

Along the boundaries, auxiliary points (*) outside Ω are introduced which are used for approximating the normal derivatives by central differences.

Let $\overset{0}{\Omega}_h$ and $\partial\Omega_h$ be the sets of gridpoints in the interior of Ω and on the boundary $\partial\Omega$, respectively. Then the discrete problem can be stated as follows:

$$\begin{aligned} &(i) \quad \Delta_h v_h = f(P) && (P\varepsilon\overset{0}{\Omega}_h) \\ &(ii) \quad \Delta_h u_h - v_h = 0 && (P\varepsilon\overset{0}{\Omega}_h \cup \partial\Omega_h) \end{aligned} \tag{2.1}$$

with boundary conditions

$$\begin{aligned} u_h(P) &= g(P) && (P\varepsilon\partial\Omega_h) \\ \frac{1}{2\Delta x}[1 \quad 0 \quad -1]u_h(P) &= \overline{g}(P) && (P=(0,y)\varepsilon\partial\Omega_h) \\ \frac{1}{2\Delta x}[-1 \quad 0 \quad 1]u_h(P) &= \overline{g}(P) && (P=(A,y)\varepsilon\partial\Omega_h) \end{aligned} \tag{2.2}$$

$$\frac{1}{2\Delta y}\begin{bmatrix}1\\0\\-1\end{bmatrix}u_h(P) = \bar{g}(P) \qquad (P=(x,B)\varepsilon\partial\Omega_h)$$

$$\frac{1}{2\Delta y}\begin{bmatrix}-1\\0\\1\end{bmatrix}u_h(P) = \bar{g}(P) \qquad (P=(x,0)\varepsilon\partial\Omega_h). \tag{2.2}$$

Here Δ_h denotes the usual second order 5-point-approximation of the Laplacian

$$\Delta_h = \frac{1}{\Delta y^2}\begin{bmatrix} & 1 & \\ q^2 & d & q^2 \\ & 1 & \end{bmatrix}$$

with $d=-2(1+q^2)$ and meshsize ratio $q=\Delta y/\Delta x$. At the corner points P, $\bar{g}(P)$ in (2.2) is to be understood as the continuous extension of $\bar{g}$ along the respective piece of $\partial\Omega$ into the corner points.

On a square grid (2.1) and (2.2) are equivalent to the 13-point-approximation (1.2) of the biharmonic problem (1.1) with quadratic extrapolation on the boundary, yielding the same discrete solution u_h, |7|.

By use of the boundary conditions (2.2), the auxiliary grid-points outside Ω can be eliminated from (2.1(ii)), of course. In the situation shown in Fig. 2.2, for example, this would give:

$$\frac{2}{\Delta y^2}u_h(P_N) - v_h(P) = -\frac{1}{\Delta y^2}[q^2 \;\; d \;\; q^2]\, g(P) - \frac{2}{\Delta y}\,\bar{g}(P)\,. \tag{2.3}$$

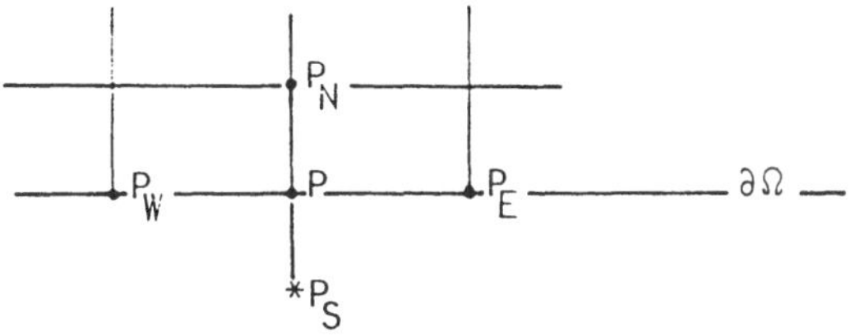

Fig. 2.2

Considered as an equation for the unknown boundary value $v_h(P)$, (2.3) is just the well known Thoms's condition.

At the corner points, one gets similarily from (2.1(ii)) and (2.2) an explicit formula for the corner values of v_h depending only on the given data g and $\bar{g}$. Although all other unknowns in (2.1) are completely decoupled from these corner values, it is convenient for the multigrid method to compute them - especially on the coarser grids where they are used for interpolation.

3. THE COMPONENTS OF THE MULTIGRID METHOD

In this section, the components of the multigrid method for solving the discrete problem (2.1), (2.2) are stated. Here, we give neither a description of the multigrid principle itself nor of its different algorithmic realisations; for this we refer to |3|, |9| using in the following the same notation and the same terminology as there.

The <u>coarsening</u> of the given finest grid, on which the problem (2.1), (2.2) is to be solved, is done in the standard way by simply doubling the meshsizes in both x- and y-directions. On all grids occuring in the multigrid process, the respective difference operators arising from the discretization described in the previous section are used. Therefore, on any grid Ω_h the discrete problem is

$$\begin{array}{llll} \Delta_h v_h(P) & = f_h(P) & (P\varepsilon\overset{0}{\Omega}_h) & \\ \Delta_h u_h(P) - v_h(P) & = \bar{f}_h(P) & (P\varepsilon\overset{0}{\Omega}_h\cup\partial\Omega_h) & (3.1) \end{array}$$

with boundary conditions of the form (2.2). On the finest grid, the right hand sides in (3.1) are those given in (2.1): $f_h(P)=f(P)$, $\bar{f}_h(P)=0$; on all coarser grids, they result from the restriction of residuals of the next finer grid. The multigrid method is performed in the <u>correction scheme</u>.

Generally the most sensitive component of a multigrid method is the relaxation used for smoothing of errors and residuals. For the biharmonic problem at hand, special care must be taken in regard to the relaxation at gridpoints near to the boundaries since no boundary conditions for the grid function v_h are provided by (2.2); the boundary values of v_h must be computed from the values of u_h at gridpoints on and near to the boundary.

We use a <u>collective relaxation scheme</u> which treats the two difference equations (3.1) in coupled form especially near to

the boundary. Its <u>pointwise</u> version is described in the following assuming the "isotropic" case $\Delta x = \Delta y =: h$:
The interior gridpoints $P \varepsilon \mathring{\Omega}_h$ are scanned in some prescribed order. Let $\overline{u}_h$ and $\overline{v}_h$ be the current approximations to u_h and v_h.

(i) At gridpoints $P \varepsilon \mathring{\Omega}_h$, without neighbouring points on $\partial\Omega_h$ (Fig. 3.1), new values $\overline{u}_h(P)$ and $\overline{v}_h(P)$ are computed by solving the 2x2-system:

$$\begin{vmatrix} -4 & 0 \\ -h^2 & -4 \end{vmatrix} \begin{vmatrix} \overline{v}_h(P) \\ \overline{u}_h(P) \end{vmatrix} = \begin{vmatrix} h^2 f_h(P) - \overline{v}_h(P_W) - \overline{v}_h(P_E) - \overline{v}_h(P_S) - \overline{v}_h(P_N) \\ h^2 \overline{f}_h(P) - \overline{u}_h(P_W) - \overline{u}_h(P_E) - \overline{u}_h(P_S) - \overline{u}_h(P_N) \end{vmatrix} . \qquad (3.2)$$

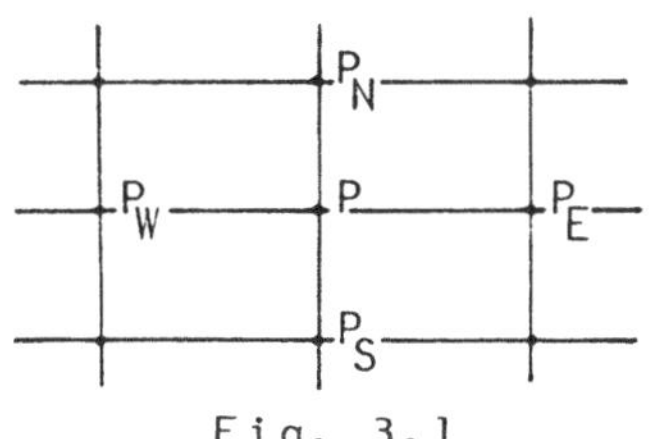

Fig. 3.1

Of course, this is equivalent to two successive Gauss-Seidel-steps for the first and the second difference equation in (3.1).

(ii) At the other gridpoints $P \varepsilon \mathring{\Omega}_h$, (3.1) involves one or two of the unknown boundary values of v_h. Here, the relaxation computes new values $\overline{u}_h(P)$ and $\overline{v}_h(P)$ simultaneously with new approximations to these boundary values by using equations of the form (2.3); note that additionally to (2.3) the inhomogeneous right hand side $\overline{f}_h$ in (3.1) must be taken into account by simply adding it to the right hand side of (2.3). If P has only one neighbouring point Q on $\partial\Omega_h$ (Fig. 3.2), this leads to a 3x3-system:

$$\begin{vmatrix} -4 & 0 & 1 \\ -h^2 & -4 & 0 \\ 0 & 2 & -h^2 \end{vmatrix} \begin{vmatrix} \overline{v}_h(P) \\ \overline{u}_h(P) \\ \overline{v}_h(Q) \end{vmatrix} = \begin{vmatrix} h^2 f_h(P) - \overline{v}_h(P_W) - \overline{v}_h(P_E) - \overline{v}_h(P_N) \\ h^2 \overline{f}_h(P) - \overline{u}_h(P_W) - \overline{u}_h(P_E) - \overline{u}_h(P_N) - g(Q) \\ h^2 \overline{f}_h(Q) - g(Q_W) - g(Q_E) + 4g(Q) - 2h\overline{g}(Q) \end{vmatrix} . \qquad (3.3)$$

If P has two neighbouring points Q^1 and Q^2 on $\partial\Omega_h$ (Fig. 3.3), one gets analogously a 4x4-system:

$$\begin{vmatrix} -4 & 0 & 1 & 1 \\ -h^2 & -4 & 0 & 0 \\ 0 & 2 & -h^2 & 0 \\ 0 & 2 & 0 & -h^2 \end{vmatrix} \begin{vmatrix} \bar{v}_h(P) \\ \bar{u}_h(P) \\ \bar{v}_h(Q^1) \\ \bar{v}_h(Q^2) \end{vmatrix} = \begin{vmatrix} h^2 f_h(P) - \bar{v}_h(P_N) - \bar{v}_h(P_E) \\ h^2 \bar{f}_h(P) - \bar{u}_h(P_N) - \bar{u}_h(P_E) - g(Q^2) - g(Q^1) \\ h^2 \bar{f}_h(Q^1) - g(C) - g(Q^1_E) + 4g(Q^1) - 2h\bar{g}(Q^1) \\ h^2 \bar{f}_h(Q^2) - g(C) - g(Q^2_N) + 4g(Q^2) - 2h\bar{g}(Q^2) \end{vmatrix} \qquad (3.4)$$

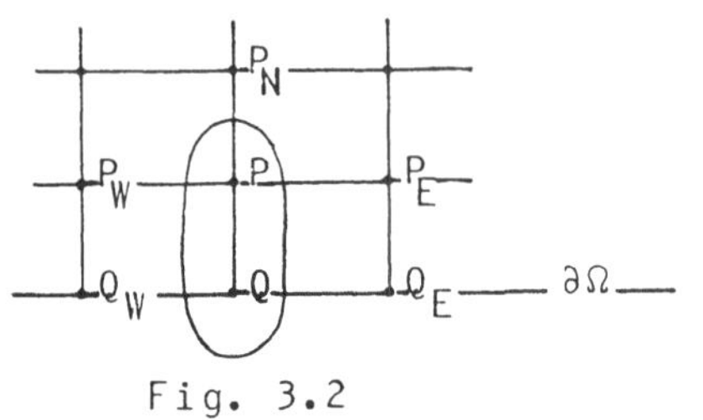

Fig. 3.2

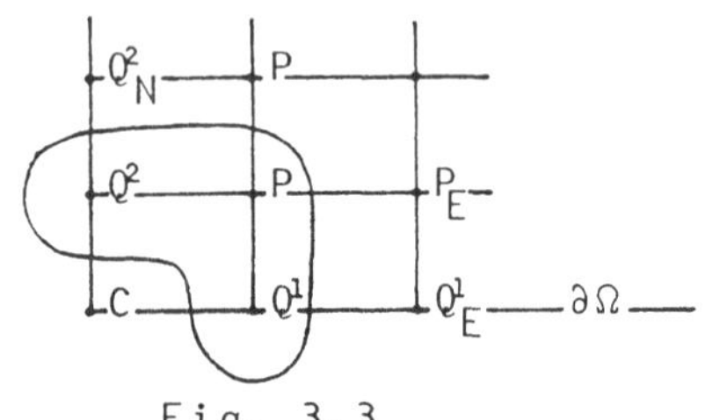

Fig. 3.3

The results presented in the next section refer to lexicographic ordering of gridpoints. With this ordering, the pointwise relaxation described above has an interior smoothing factor $\mu=0.5$ for $\Delta x=\Delta y$.[1] It gets worse in the "anisotropic" case $\Delta x\neq\Delta y$. In this case line-relaxation must be used (columnwise if $q\leq 1$, rowwise if $q\geq 1$). We do it also collectively, i.e. all values of $\bar{u}_h$ and $\bar{v}_h$ belonging to a gridline are changed simultaneously using again equations of the form (2.3) for the determination of the unknown boundary values of v_h. This results in 5-diagonal systems per gridline which are solved by Gaussian elimination; for more details we refer to |7|. Using again the lexicographic ordering for the gridlines, the smoothing factor of the line-relaxation is $\mu=0.447$. The numerical results in section 4 show, that the estimate $\rho\simeq\mu^{\nu}$ (for small ν, ν being the total number of relaxations on the finest level) of the asymptotic convergence factor ρ of the multigrid iteration is in fact reached for the lexicographic relaxations using W-cycles. Compared to the lexicographic versions, due to some boundary effects checkered and zebra-relaxations do not give any improvement although having the better interior smoothing factors. This phenomenon is not yet investigated in detail. Numerical

[1] Note, that local mode analysis for the system (3.1) yields exactly the same results as in case of the 5-point-approximation of Poisson's equation since the coupling of (3.1) by the boundary conditions is ignored in this analysis. A description of local mode analysis can be found in |3|, |9|.

results concerning checkered and zebra-relaxation are shown in |7|.

After having smoothed out the error on a fine grid Ω_h in the manner described above, the values of $\bar{u}_h$ at the auxiliary grid-points outside Ω may be chosen such that the discrete boundary conditions are fulfilled. Then, the coarse grid correction problem on Ω_{2h} has the form (3.1) (replacing h by 2h) with homogeneous boundary conditions (2.2) ($g=\bar{g}=0$). Its right hand sides are:

$$f_{2h}(P) = I_h^{2h}(f_h-\Delta_h\bar{v}_h)(P) \qquad (P\varepsilon\overset{0}{\Omega}_{2h})$$
$$\bar{f}_{2h}(P) = I_h^{2h}(\bar{f}_h-\Delta_h\bar{u}_h+\bar{v}_h)(P) \quad (P\varepsilon\overset{0}{\Omega}_{2h}\cup\partial\Omega_{2h}),$$

with the current approximations $\bar{u}_h$, $\bar{v}_h$ on Ω_h.
For this fine-to-coarse transfer I_h^{2h} of residuals, we use the normal full-weighting scheme |3|,|9| which is given by the 9-point-operator

$$\frac{1}{16}\begin{bmatrix} 1 & 2 & 1 \\ 2 & \underline{4} & 2 \\ 1 & 2 & 1 \end{bmatrix}$$

in the interior and by the 6- or 4-point-operators

$$\frac{1}{8}\begin{bmatrix} 1 & 2 & 1 \\ 1 & \underline{2} & 1 \end{bmatrix} \qquad \text{or} \qquad \frac{1}{4}\begin{bmatrix} 1 & 1 \\ \underline{1} & 1 \end{bmatrix}$$

on the boundary or at the corner points, respectively. Note, that after the relaxation, the residuals of the second difference equation in (3.1) vanish at all boundary gridpoints.

The coarse-to-fine transfer of corrections in the multigrid cycle is performed by bilinear interpolation.
In the full-multigrid mode, the approximate solutions computed on a coarse grid are prolonged to the next finer grid by bicubic interpolation. In |7| also a more sophisticated technique for this FMG-interpolation is described using difference operators and intermediate grids which gives still slightly better results.

4. NUMERICAL RESULTS

In this section we give numerical results which refer to the multigrid iteration (MGI) and to the full-multigrid method (FMG). In both variants we use the W-cycle, |9|. Numerical results concerning other types of cycles can be found in |7|.

On the actual coarsest grid in the multigrid cycle, the corresponding correction problem is solved nearly exactly. Usually, however, only few relaxation sweeps (5-8) on the coarsest grid provide a sufficient accuracy.

Table 1 and Table 2 show asymptotic convergence factors of the multigrid iteration. These were computed by a v. Mises-vector-iteration. ν_1 and ν_2 are the numbers of smoothing steps before and after the coarse grid correction, respectively.
In Table 1, the isotropic case is considered (meshsize ratio q=1). Here the pointwise lexicographic relaxation described in Section 3 is used for smoothing. The domain Ω is the unit square. N denotes the number of gridpoints in both x- and y-directions, yielding $\Delta x=\Delta y=1/N$.
The total number of grids in the multigrid cycle are given in the second column of Table 1. The coarsest grid always has the meshsizes $\Delta x=\Delta y=1/4$. In the last row of Table 1, the estimates μ^ν for the convergence factors resulting from local mode analysis are stated (μ being the smoothing factor and $\nu=\nu_1+\nu_2$ the number of smoothing steps on the finest grid).

Table 1: Convergence factors of multigrid iteration

		ν_1,ν_2				
N	grids	1,0	2,0	3,0	1,1	2,1
16	3	0.324	0.192	0.085	0.182	0.086
32	4	0.373	0.203	0.091	0.206	0.084
64	5	0.388	0.210	0.092	0.214	0.092
128	6	0.403	0.213	0.108	0.216	0.108
256	7	0.402	0.214	0.119	0.217	0.117
μ^ν		0.500	0.250	0.125	0.250	0.125

Table 2 summarizes results for some anisotropic cases. The domain Ω is the rectangle $[0,A]\times[0,1]$. The actual finest grid has the meshsizes $\Delta x = A/N$ and $\Delta y = 1/N$ with meshsize ratio $q = 1/A$. Lexicographic line-relaxation is used for smoothing.

Table 2: Convergence factors of multigrid iteration

	N / grids / A	16 / 3	32 / 4	64 / 5	128 / 6	256 / 7	512 / 8	μ^{ν}
$\nu_1=1$, $\nu_2=1$	2	0.069	0.100	0.115	0.131	0.139	0.135	0.200
	4	0.046	0.079	0.095	0.117	0.139	0.134	
	8	0.007	0.051	0.081	0.100	0.135	0.134	
	16	<0.001	0.007	0.054	0.092	0.130	0.132	
	32	<0.001	<0.001	0.007	0.064	0.113	0.128	
$\nu_1=2$, $\nu_2=1$	2	0.039	0.043	0.061	0.071	0.080	0.088	0.089
	4	0.016	0.043	0.042	0.059	0.074	0.088	
	8	0.001	0.018	0.043	0.043	0.064	0.087	
	16	<0.001	0.001	0.019	0.045	0.048	0.086	
	32	<0.001	<0.001	0.001	0.021	0.055	0.083	

The results in Table 1 and 2 show that the convergence factors are in fact asymptotically bounded by the theoretical values μ^{ν}; for small ν they are partly even below these values. Of course, the very good convergence factors on the coarse grids are an effect of relaxation which converges still rather fast on these grids.
We point out further that the convergence factors shown are comparable to those of multigrid iterations for the Poisson-equation using lexicographic relaxations for smoothing, |9|.

The numbers of arithmetic operations per gridpoint of the finest grid required for one multigrid cycle are

$$18\nu+38 \text{ add./sub. and } 6\nu+11 \text{ mult.}$$

in the isotropic case (pointwise relaxation) and

$$24\nu+38 \text{ add./sub. and } 20\nu+15 \text{ mult.}$$

in the anisotropic case (line-relaxation). Compared to the

corresponding multigrid cycles for the Poisson equation this computational work is higher by a factor between 2.3 and 3.

Table 3 contains typical results of the full-multigrid method. The given values are errors in the maximum- and discrete L_2-norm ($\|\cdot\|_\infty$ and $\|\cdot\|_2$) both taken over all gridpoints without the corner points. They refer to the test example with continuous solutions $u=x^3\ln(1+y)+y/(1+x)$, $v=\Delta u$ on the unit square, which was also used in |1|.

u_h and v_h are the respective exact solutions; $\tilde{u}_h$ and $\tilde{v}_h$ denote the approximate solutions computed by FMG. As in Table 1, N is the parameter of the finest grid in the multigrid process with meshsizes $\Delta x=\Delta y=1/N$. The number of grids used in the multigrid correction cycles are also the same as in Table 1. Again the pointwise lexicographic relaxation is used for smoothing.
The full-multigrid method starts on the second grid (with meshsizes $\Delta x=\Delta y=1/8$). The discrete problem there is solved approximately by a fixed number of two-grid iterations (here: 5). On each finer level only one multigrid cycle is performed.

Table 3: FMG-results

			$\nu_1=2, \nu_2=1$		$\nu_1=3, \nu_2=0$		
N	$\|u-u_h\|$	$\|v-v_h\|$	$\|\tilde{u}_h-u_h\|$	$\|\tilde{v}_h-v_h\|$	$\|\tilde{u}_h-u_h\|$	$\|\tilde{v}_h-v_h\|$	
16	.146(-3)	.542(-1)	.858(-4)	.229(-1)	.707(-4)	.732(-2)	$\|\cdot\|_\infty$
32	.369(-4)	.275(-1)	.198(-4)	.144(-1)	.102(-4)	.306(-2)	$\|\cdot\|_\infty$
64	.928(-5)	.138(-1)	.353(-5)	.826(-2)	.247(-5)	.157(-2)	$\|\cdot\|_\infty$
128	.232(-5)	.693(-2)	.823(-6)	.439(-2)	.424(-6)	.799(-3)	$\|\cdot\|_\infty$
256	.581(-6)	.347(-2)	.202(-6)	.226(-2)	.147(-6)	.399(-3)	$\|\cdot\|_\infty$
16	.567(-4)	.726(-2)	.324(-4)	.444(-2)	.249(-4)	.154(-2)	$\|\cdot\|_2$
32	.143(-4)	.205(-2)	.677(-5)	.186(-2)	.307(-5)	.403(-3)	$\|\cdot\|_2$
64	.359(-5)	.564(-3)	.150(-5)	<u>.769(-3)</u>	.118(-5)	.154(-3)	$\|\cdot\|_2$
128	.898(-6)	.153(-3)	.345(-6)	<u>.297(-3)</u>	.160(-6)	.484(-4)	$\|\cdot\|_2$
256	.225(-6)	.407(-4)	.864(-7)	<u>.110(-3)</u>	.664(-7)	.197(-4)	$\|\cdot\|_2$

In the maximum-norm the FMG-errors $\tilde{u}_h-u_h$ and $\tilde{v}_h-v_h$ are always below the global discretization errors $u-u_h$ and $v-v_h$. In the L_2-norm this is not achieved for the v-errors on fine grids in

cases in which relaxations are performed after the coarse grid correction ($\nu_2 \neq 0$), (see the underlined values in Table 3).
A more detailed numerical investigation shows that these relaxations slightly reduce the error in u on one hand, but, on the other hand, the error in v is enlarged especially on and near to the boundary, |7|.
In all the tests we made, the choice $\nu_1=3$, $\nu_2=0$, however, provided reliable FMG-solutions of discretization accuracy; usually even $\nu_1=2$, $\nu_2=0$ was sufficient.

The asymptotic computational effort per gridpoint of the finest grid for the full-multigrid method with only one W-cycle per level is

$$\frac{4}{3}(18\nu+38)+6 \text{ add./sub.} \quad \text{and} \quad \frac{4}{3}(6\nu+11)+4 \text{ mult.}$$

in the isotropic case and

$$\frac{4}{3}(24\nu+38)+6 \text{ add./sub.} \quad \text{and} \quad \frac{4}{3}(20\nu+15)+4 \text{ mult.}$$

in the anisotropic case.
Computing times (in seconds) for the full-multigrid method are shown in Table 4. They refer to the method used in Table 3 (pointwise relaxation) and were measured on a IBM 3083 B (compiler BS 3000, Opt=3). All arithmetic operations were performed in double precision.

Table 4: FMG computing times

N	$\nu_1=2,\ \nu_2=0$	$\nu_1=3,\ \nu_2=0$
16	0.05	0.06
32	0.12	0.14
64	0.39	0.46
128	1.39	1.69
256	5.34	6.50

Finally we would like to point out that the multigrid techniques presented here are not claimed to be optimal. Certainly they can be improved - especially with regard to efficiency. In the form and with the components and parameters used above (W-cycle, 2-3 relaxations per level), the full - multigrid method for the biharmonic equation is not believed to be able to compete with the fastest method for this problem, which is available today and which was proposed by P. Bjorstad |1|, although having also the optimal complexity $O(N^2)$. Without having made precise comparisons, we assume that Bjorstad's method is faster by a factor between 1.5 and 2. We see the main advantage of our multigrid techniques in the fact, that they are directly applicable to more general situations including variable coefficients and non-linearities (e.g. |7|).

REFERENCES

|1| Bjørstad, P.: Numerical solution of the biharmonic equation. Ph.D. Thesis, Dept. of Computer Science, Stanford University, 1981.

|2| Bramble, J.H.: A second order finite difference analog of the first biharmonic boundary value problem. Numerische Mathematik, 9, 1966.

|3| Brandt, A.: Multigrid techniques: 1984 guide with applications to fluid dynamics. GMD-Studie No. 85, Gesellschaft für Mathematik und Datenverarbeitung, St. Augustin, 1984.

|4| Buzbee, B.L., Dorr, F.W.: The direct solution of the biharmonic equation on rectangular regions and the Poisson equation on irregular regions. SIAM J. Numer. Anal., Vol. 11, 1974.

|5| Holstein, H.: A multigrid treatment of Stokesian boundary conditions. Lecture, held at the Oberwolfach-conference on Multigrid Methods (9.12.-15.12.1984).

|6| Kuttler, J.R.: A finite difference approximation for the eigenvalues of the clamped plate. Numerische Mathematik, 17, 1971.

|7| Linden, J.: Dissertation, Universität Bonn, to appear.

|8| Papamanolis, G.: Multigrid methods in fluid dynamics. Master Thesis, Dept. of Computer Science, University of Wales, Aberystwyth, 1984.

|9| Stüben, K., Trottenberg, U.: Multigrid methods: fundamental algorithms, model problem analysis and applications. Proceedings of the conference on Multigrid Methods, Köln 1981 (W.Hackbusch, U.Trottenberg, eds.). Lecture Notes in Mathematics, 960, Springer Verlag, Berlin, 1982.

|10| Vajtersic, M.: A fast algorithm for solving the first biharmonic boundary value problem. Computing 23, 1979.

|11| Zlamal, M.: Discretization and error estimates for elliptic boundary value problems of the fourth order. SIAM J. Numer. Anal., Vol. 4, 1967.

A FAST SOLVER FOR THE STOKES EQUATIONS USING MULTIGRID WITH A UZAWA SMOOTHER

J.F. Maitre, F. Musy, P. Nigon

Département de Mathématiques-Informatique-Systèmes

Ecole Centrale de Lyon, BP 163

F 69131 ECULLY CEDEX France

SUMMARY

We propose a smoother of Uzawa type suitable for the solution of mixed problems by multigrid methods. We report numerical experiments for the Stokes equations discretized through the M.A.C. scheme, comparing the computational work with that of the distributive relaxation.

1. LINEAR SYSTEMS OF MIXED TYPE, THE CASE OF THE STOKES EQUATIONS

By "mixed systems" we mean systems of the following type :

$$L_h X_h = F_h \,, \tag{1}$$

where $X_h = (U,P)$, $F_h = (b,c)$ are in $R^{k_h} \times R^{m_h}$ and L_h has the structure

$$L_h = \begin{bmatrix} A & B^t \\ B & 0 \end{bmatrix} \,. \tag{2}$$

A system such as (1), (2) may be obtained by a finite element discretization of a mixed variational problem :

$$\left.\begin{array}{ll} (u,\lambda) \in V \times M, & \\ a(u,v) + b(v,\lambda) = (f,v), & \forall\, v \in V, \\ b(u,\mu) \qquad\quad\;\; = (\phi,\mu), & \forall\, \mu \in M. \end{array}\right\} \tag{3}$$

In this case, under conditions including the L.B.B. one (see for instance [2]), A is symmetric positive definite, B^t injective and thus L_h regular.

If A is positive definite and B^t is not an injection, (1) has a solution if c is in R(B), the range of B, (compatibility condition) ; U is unique whereas P is not and we denote $\tilde{P}$ the unique solution of minimal norm which is in R(B).

We consider the Stokes equations in the domaine Ω with Dirichlet condition on $\Gamma = \partial\Omega$:

$$\left.\begin{array}{r} -\Delta\vec{u} + \nabla p = \vec{f} \\ \operatorname{div} \vec{u} = 0 \\ \vec{u} = \vec{u}_\Gamma \end{array}\right\} \tag{4}$$

assuming that the compatibility condition $\int_\Gamma \vec{u}_\Gamma . \vec{\nu}\, d\Gamma = 0$ is satisfied.

The problem (4) admits formulations of type (3) and various finite element discretizations (see for instance [9], [10]), but here we restrict us to the direct finite difference discretization through the M.A.C. scheme [12] of the 2-D problem in $\Omega =]0,1[^2$. The domain Ω is divided in $(n+1)^2$ square cells of side $h = 1/(n+1)$; the discrete velocity u_h (resp. v_h) is defined at the midpoints of the vertical (resp. horizontal) edges, and the discrete pressure at the cell centers ("staggered grid"). The operators Δ and ∇ are discretized as usually, leading to the equations

$$\begin{aligned}\frac{1}{h^2}\,[4u_h(x,y) - u_h(x-h,y) - u_h(x+h,y) - u_h(x,y-h) - u_h(x,y+h)] \\ + \frac{1}{h}\,[p_h(x+\tfrac{h}{2},y) - p_h(x-\tfrac{h}{2},y)] = f_1(x,y)\end{aligned} \tag{5}$$

at the u_h-nodes (with some changes near the boundary), equations of the same type at the v_h-nodes, and to the equations

$$u_h(x+\tfrac{h}{2}, y) - u_h(x-\tfrac{h}{2}, y) + v_h(x,y+\tfrac{h}{2}) - v_h(x,y-\tfrac{h}{2}) = 0 \quad (6)$$

at the p_h-nodes. For the lexicographic ordering of the three groups of unknowns, it is easy to write down the matrices A,B, and to obtain a system of type (1), (2), with $k_h = 2n(n+1)$, $m_h = (n+1)^2$. Here c is in R(B) iff c has zero mean value ; to ensure this discretized compatibility condition, the right hand side 0 of (6) must be replaced by the constant

$$\begin{aligned}\bar{c} = h^2 \sum_{j=1}^{n+1} [u_\Gamma(1,(j-\tfrac{1}{2})h) + v_\Gamma((j-\tfrac{1}{2})h,1) \\ - u_\Gamma(0,(j-1/2)h) - v_\Gamma((j-1/2)h,0)].\end{aligned}$$

The minimal norm solution $\tilde{P}$ is the projection onto R(B) of any solution P, that is $\tilde{P} = P - \bar{p}$ where $\bar{p}$ is the vector with all components equal to the mean value of the vector P.

We notice for the sequel that for this problem :

$$A = \begin{bmatrix} A_1 & 0 \\ 0 & A_2 \end{bmatrix}, \quad B = (B_1, B_2), \quad b = \begin{bmatrix} b_1 \\ b_2 \end{bmatrix}, \quad U = \begin{bmatrix} u \\ v \end{bmatrix}. \quad (7)$$

2. MULTIGRID COMPONENTS FOR MIXED PROBLEMS AND STOKES EQUATIONS

2.1. The Uzawa type smoothers

For the general mixed system (1), (2), a smoother using a Richardson type iteration applied to the "normal equations" has been introduced and studied by W. Hackbusch [5] (see also [8], [11]). It is defined by

$$X_h \leftarrow X_h - \omega \begin{bmatrix} \tilde{A}^t & \tilde{B}^t \\ \tilde{B} & 0 \end{bmatrix} (L_h X_h - F_h),$$

where $\tilde{A} = h^{2\alpha} A$, $\tilde{B} = h^{2\beta} B$; α,β denoting the orders of the differential operators.

For the particular system of the discretized Stokes equations (5), (6), A. Brandt and N. Dinar [1], [3], have introduced the so called distributive relaxation. This smoother consists in one step of Gauss-Seidel for the equations of type (5), followed by one distributive relaxation which, for each equation of type (6), changes the 4 values of velocity appearing in (6) and the 5 neighbouring values of the pressure, in order to verify the equation (6) while keeping unchanged the residuals of equations (5).

We propose, for the general mixed system (1), (2), a smoother based on the following Uzawa iteration in its "augmented lagrangian" form (see [4]) :

$$\left.\begin{aligned} (A + r\,B^tB)\,U^{n+1} &= b - B^tP^n + r\,B^tc\,, \\ P^{n+1} &= P^n + \rho\,(BU^{n+1} - c)\,, \end{aligned}\right\} \qquad (8)$$

where $r \geqslant 0$, $\rho > 0$ are two parameters. For this smoother, the exact solution U^{n+1} in (8) is only approximated by a few steps of the S.O.R. method (parameter ω) starting from U^n.

For the exact Uzawa smoother (8) with $r = 0$, we refer to [6] for a theoretical study in a variational framework.

For systems with structure (7), we propose the following smoother which has been tested on the Stokes system. It transforms $(\mathring{u},\mathring{v},\mathring{p})$ into $(\bar{u},\bar{v},\bar{p})$ as follows :

- starting from $\mathring{u}$, $\tilde{u}$ results from one step of S.O.R. on

$$(A_1 + r\,B_1^t\,B_1)\,u = b_1 - B_1^t\,\mathring{p} + r\,B_1^t\,c - r\,B_1^t\,B_2\,\mathring{v},$$

- starting from $\mathring{v}$, $\bar{v}$ results from two steps of S.O.R. on

$$(A_2 + r\,B_2^t\,B_2)\,v = b_2 - B_2^t\,\mathring{p} + r\,B_2^t\,c - r\,B_2^t\,B_1\,\tilde{u}, \qquad (9)$$

- starting from $\tilde{u}$, $\bar{u}$ results from one step of S.O.R. on

$$(A_1 + r\,B_1^t\,B_1)\,u = b_1 - B^t\,\mathring{p} + r\,B_1^t\,c - r\,B_1^t\,B_2\,\bar{v},$$

- $\bar{p} = \mathring{p} + \rho\,(B_1\,\bar{u} + B_2\,\bar{v} - c)$.

This smoother will be denoted UZAL if $r \neq 0$ (Uzawa - augmented lagrangian) and UZA if $r = 0$. In the latter case, the first 3 steps in (9) simply consists in 2 steps of S.O.R. on $AU = b - B^tp$.

2.2. The other Multigrid components

The restriction and prolongation operators for a mixed system have two components :

$$\left.\begin{aligned} &\text{restriction } r_\ell\,(b_\ell, c_\ell) = (r_\ell^1\,b_\ell,\, r_\ell^2\,c_\ell) \\ &\text{prolongation } p_\ell(U_{\ell-1}, P_{\ell-1}) = (p_\ell^1\,U_{\ell-1},\, p_\ell^2\,P_{\ell-1}) \end{aligned}\right\} \qquad (10)$$

and the matrix at level ℓ has the form

$$L_\ell = \begin{bmatrix} A_\ell & B_\ell^t \\ B_\ell & 0 \end{bmatrix} . \qquad (11)$$

If B^t is not an injection, the compatibility condition must be satisfied at each level, and it is advised to project on $R(B_\ell)$ after each coarse-to-fine transfer.

For the Stokes equations discretized as in §1, the coarse-to-fine transfer of u,v,p are defined by means of the bilinear interpolation using the 4 nearest nodes of the coarse grid. At the coarse u (resp. v)-nodes, the restricted defect is taken to be the mean value of the defects associated to the 2 nearest u (resp. v) nodes in the fine grid. At the coarse p-nodes, the restricted defect is the mean value of the defects associated to the 4 nearest nodes in the fine grid (for details see [1]). With such a restriction for the defects at p-nodes, the compatibility condition is automatically satisfied on all levels provided it holds on the finest one.

3. MULTIGRID SOLUTION OF STOKES EQUATIONS WITH UZAWA TYPE SMOOTHERS

All the numerical experiments reported here correspond to the solution of the bidimensional Stokes equations (4) with solution $u(x,y) = - v(x,y) = \sin 3\,(x+y), p = - 6 \cos 3\,(x+y) + c^{te}$, that is with $f_1(x,y) = 36 \sin 3\,(x,y)$, $f_2(x,y) = 0$, $u_\Gamma(x,y) = \sin 3\,(x+y)$, $v_\Gamma(x,y) = - \sin 3(x+y)$, with the discretization described in §1, and the MG components detailed in §2.

The smoothers UZAL and UZA are tested and some comparisons made with the distributive relaxation.

The convergence rate σ is estimated for i steps by

$$\sigma(i) = (R(i) \;/\; R(o))^{1/i}, \qquad (12)$$

where $R(i) = ||A_1 u_h^{(i)} + B_1^t\, p_h^{(i)} - b_1|| + ||A_2\, v_h^{(i)} + B_2^t p_h^{(i)} - b_2|| +$

$+ ||B_1\, u_h^{(i)} + B_2\, v_h^{(i)} - c||$, $||.||$ denoting the euclidean norm.

3.1. Two-grid convergence rates ; choice of the smoothing parameters

The two levels correspond here to $h = 1/32$, $H = 1/16$ (3008 and 736 unknowns).

The best choice for ω in the S.O.R. steps of (9) is 1, both for convervenge rate and computational work. For example for the two-grid method with $\nu = 2$ steps of the smoother UZAL ($r = 1$, $\rho = 1.5$), the convergence rate, defined by (12), with respects to ω is given by :

ω	0.8	1	1.2	1.4
$\sigma(10)$	0.163	0.151	0.240	0.484

For a fixed number ν of smoothing steps of UZAL, there is an optimal choice for (r,ρ) denoted (r_o,ρ_o). For the efficiency UZA is to prefer, since the computational work is more than twice for UZAL $(r \neq 0)$.

The following table gives, with respects to ν, the convergence rates $\sigma(10)$ of the three smoothers distributive relaxation, UZA(ρ_o), UZAL(r_o,ρ_o), all of them with $\omega = 1$:

smoother \ ν	1	2	3	4
distr. relax.	0.378	0.247	0.171	0.131
UZA	0.358	0.181	0.115	0.088
UZAL	0.287	0.146	0.082	0.058

3.2. Multigrid and Full Multigrid results

The coarsest grid is chosen at h = 1/4 (40 unknowns) for all the following tests.

About the only parameter ρ of the smoother UZA, the best choice ρ_o for a V-cycle with 2 smoothings before is given for a sequence of finest levels in the following table :

h	1/8	1/16	1/32	1/64	1/128
ρ_o	1.09	0.98	0.85	0.44	0.34

However the convergence rate is not very sensitive to the choice of ρ ; in fact the simple choice $\rho = 1$ is not so bad (for instance the efficiency for h = 1/64 is nearly the one obtained with $\rho = \rho_o = 0.44$).

To show up the efficiency of the UZA smoother and compare it with the distributive relaxation one, we give in the following table the computational work corresponding to the optimal strategy with respects to the number N of unknowns, this for each h (finest level) and both for MG and FMG (with one inner step of M G at each level). We make precise that the stopping criterion is for all these tests "e(i) sufficiently close to e", where these quantities are defined by [11]

$$e^2(i) = ||u - u_h^i||^2 + ||v - v_h^i||^2 + h^2|| p - p_h^i||^2,$$

$$e^2 = ||u - u_h||^2 + ||v - v_h||^2 + h^2 ||p - p_h||^2,$$

$||.||$ denoting the discrete L^2-norm.

h	1/8	1/16	1/32	1/64	1/128	1/256
N	176	736	3008	12160	48896	196096
M G distr. relax.	68 N Log N	57 N Log N	55 N Log N	56 N Log N	57 N Log N	
M G UZA	52 N Log N	48 N Log N	45 N Log N	48 N Log N	55 N Log N	
FMG distr. relax.	314 N	277 N	251 N	256 N	215 N	
FMG UZA	230 N	255 N	222 N	131 N	132 N	132 N

CONCLUDING REMARKS

The experiments reported here belong to the thesis of P. Nigon [7], where can be found more details. Smoother UZA presented here appears to be very promising. Its efficiency compares favourably with that of the distributive relaxation. Moreover it may be used for any mixed systems, in particular for finite element discretizations on general meshes of the Stokes equations, about which we have work in progress.

REFERENCES

[1] BRANDT,A., DINAR,N. : "Multigrid solutions to elliptic flow problems", Numerical Methods for PDES, S.V. Parter ed., pp. 53-147, Academic Press, New York (1979).

[2] BREZZI,F. : "On the existence, uniqueness and approximation of Saddle-point problems arising from Lagrangian multipliers", RAIRO, Anal. Numer. 8 (R-2), pp. 129-151 (1974).

[3] DINAR,N. : "Fast methods for the numerical solution of boundary value problems", Ph. D. Thesis, Dept. of Applied Mathematics, Weizmann Institute of Science, Rehovot (1978).

[4] FORTIN,M., GLOWINSKI,R. : "Méthodes de lagrangien augmenté (Application à la résolution numérique de problèmes aux limites)", Collection MMI, Dunod, Paris (1981).

[5] HACKBUSCH, W. : "Analysis and multigrid solutions of mixed finite element and mixed difference equations", Preprint, Institut für angewandte Mathematik, Ruhr-Universität Bochum (1980).

[6] MUSY, F. : Thèse, Université de Lyon, to appear.

[7] NIGON,P. : "Une nouvelle classe de méthodes multigrilles pour les problèmes mixtes - Application à la résolution rapide du problème de Stokes", Thèse de 3e Cycle, Ecole Centrale de Lyon (1984).

[8] PEISKER,P. : "A multilevel algorithm for the biharmonic problem", Bericht Nr. 17, Abteilung für Mathematik, Ruhr-Universität Bochum (1984). Numer. Math. (to appear)

[9] RAVIART,P.A. : "Les méthodes d'éléments finis en mécanique des fluides", Ecole d'Eté d'Analyse Numérique CEA-EDF-INRIA, Eyrolles, Paris (1981).

[10] TEMAM,R. : "Navier-Stokes equations, theory and numerical analysis", North-Holland, revised ed. (1979).

[11] VERFÜRTH,R. : "A multilevel algorithm for mixed problems", SIAM J. Numer. Anal. 4, 441-455 (1984)

[12] WELCH,J.E., HARLON,F.H., SHANNON,J.P., DALY,B.J. : "The MAC method", LASL, Report N° LA 3425, Los Alamos Scientific Laboratory (1965).

CALCULATIONS OF TRANSONIC FLOWS AROUND SINGLE AND MULTI-ELEMENT AIRFOILS ON A SMALL COMPUTER

Zenon P. Nowak
Warsaw University of Technology
00-665 Warszawa, Nowowiejska 24

1. SUMMARY

The aim of this paper is to present a method for calculating the transonic flows around the single and multi-element airfoils. The main characteristic of the present method is the minimum possible storage requirement of one real per node of the finite difference grid.

The method is based on the finite volume discretization of the full potential equation with artificial compressibility [5]. The velocity potential is calculated on a succession of grids by the line-Newton method. The successively finer grids are obtained from the coarser ones by adding every second grid-line, enclosing the airfoil. The coarse-grid results are prolongated to the finer grids by solving the independent nonlinear systems of the mass-conservation equations for the unknown values of the potential at the points of the added grid-lines. The same treatment is given to all the closed lines during the subsequent iteration process. Each nonlinear system is solved approximately by a single iteration of the Newton method, overwriting the previous values of the potential on a current line. The Newton-linearized systems are led to the three-diagonal matrix form and solved exactly by the method of recursive cyclic reduction [4]. To save storage, the coordinates of the grid-points are recomputed whenever necessary from the simple semi-analytic formulas.

In the case of the multi-element airfoils, the channels between the component parts are treated with the similar line-Newton relaxation or underrelaxation procedure. The fluxes through the boundaries of the channel regions are adjusted at the beginning of each global iteration.

As shown by the results of the calculations in various cases of the transonic flows around single and multi-element airfoils, the present method is quite efficient for relatively coarse grids of the size 21 × 64 or 17 × 64. For finer grids of an examplary size 41 × 128 the convergence of the present procedure could be improved by the FAS multigrid method [2], [3].

The paper presents also an efficient method for generating nearly orthogonal grids in double-connected regions with the aid of the simple semi-analytic formulas. The resulting grids are strictly orthogonal at the inner boundary.

2. INTRODUCTION

The present work is, in many respects, a continuation of [7] and [8]. As in these earlier papers, the solution method to be presented here belongs to the family of quasi-Newton methods. In fact, it can be derived from the exact Newton method by neglecting the elements of the Jacobian matrix, which correspond to the pairs of the grid-points on the different grid-lines around the airfoil. Such a simplification, when combined with an improvement of the grid generation procedure, allows for a drastic reduction of the computer storage requirements.

The calculations for a variety of transonic flow cases were made on a small computer, available to the present author at the Institute of the Aircraft Technology and the Applied Mechanics, Warsaw University of Technology. For single airfoils the method was further tested for finer grids during a later visit of the present author at the Department of Mathematics and Informatics, Delft University of Technology.

3. THE COMPUTATIONAL METHOD FOR A SINGLE AIRFOIL

3.1 Discretization of the problem

The discrete system for a single airfoil is similar to that presented in [5]. A curvilinear grid is generated inside a truncated region around the airfoil on the (x,y) - plane as the image of the regular grid of the unit mesh size on a rectangle $0 \leq \xi \leq m$, $0 \leq \eta \leq n$ (a part of an exemplary grid is shown in Fig. 8). The correspondence of the boundaries and the grids on both planes (x,y) and (ξ,η) is shown in Figs. 1a, b. The image grid on the (x,y) plane, obtained by the procedure of Section 5, is orthogonal at the points of the airfoil surface, corresponding to the line $\eta=0$.

The flow problem is reformulated as a boundary value problem on the rectangle. The rectangle is divided into the interior and the boundary cells, such as those represented by broken lines in Fig. 1 b. Extending the grid in the ξ-direction we can adopt the same forms of the control-cells along the vertical lines $\xi=0$ and $\xi=m$.

Integrating the transformed full potential equation (see e.g., [5]) over the areas of the interior and the boundary cells we obtain the mass-conservation equations of the form:

$$(\tilde{\rho}U/I)_a - (\tilde{\rho}U/I)_c + (\rho V/I)_b - (\rho V/I)_d = 0, \tag{1a}$$

and, respectively,

$$\frac{1}{2}(\tilde{\rho}U/I)_a - \frac{1}{2}(\tilde{\rho}U/I)_c + (\rho V/I)_b = 0, \tag{1b}$$

whose component terms, calculated at the middle points of the cell faces, contain ρ: the density, U and V: the contravariant velocity components in the ξ and η-directions, respectively, I: the Jacobian of the transformation $(\xi,\eta)\to(x;y)$, $\tilde{\rho}$: the "retarded" density. The density, velocity and the space coordinates x, y are nondimensionalized by the stagnation density, the critical sound speed and the airfoil chord length, respectively.

The quantities appearing in (1a,b) follow from the formulas:

$$\rho = \left(1 - \frac{\kappa-1}{\kappa+1}\lambda^2\right)^{1/(\kappa-1)} \qquad (\kappa\text{: the adiabatic exponent}),$$

$$\lambda = |U\Phi_\xi + V\Phi_\eta|^{1/2} \qquad (\lambda\text{: the velocity modulus}),$$

$$U = A_1\Phi_\xi + A_2\Phi_\eta \qquad (\Phi\text{: the velocity potential}),$$

$$V = A_2\Phi_\xi + A_3\Phi_\eta,$$

$$A_1 = I^2(x_\eta^2 + y_\eta^2), \quad A_2 = -I^2(x_\xi x_\eta + y_\xi y_\eta),$$

$$A_3 = I^2(x_\xi^2 + y_\xi^2), \quad I = 1/(x_\xi x_\eta - y_\xi y_\eta),$$

$$\tilde{\rho} = \rho - \nu(M^2)\rho_{\overleftarrow{\xi}} \qquad (\rho_{\overleftarrow{\xi}}\text{: the upwind derivative}),$$

$$M^2 = 2\lambda^2/(\kappa + 1 - \lambda^2(\kappa - 1)) \qquad (M\text{: the local Mach number}),$$

$\nu(M^2)$: a certain smooth function increasing from zero for $M<M_l$ to the unity for $M>M_u$, where $M_l=1-\varepsilon$, $M_u=1+\varepsilon$ and ε is a small number.

The retarded densities $\tilde{\rho}_a$ and $\tilde{\rho}_c$ are calculated using the upwind grid-node values of the switching function (the values $\nu(M_5^2)$ and $\nu(M_4^2)$, respectively, when $U_a>0$, $U_c>0$; otherwise: $\nu(M_6^2)$ and $\nu(M_5^2)$).

The derivatives Φ_ξ and Φ_η at the middle points a, b, c, d of the interior cell faces are approximated to the second order accuracy with the aid of the surrounding grid-node values of the potential (e.g., for the point a these values are: Φ_2, Φ_3, Φ_5, Φ_6, Φ_8 and Φ_9).

The equations (1a) for the interior cells always contain

the unknown values $\Phi_1, \Phi_2, \ldots, \Phi_9$ of the potential at the corresponding grid-nodes in Fig. 1 b. The additional values at the nodes 1', 4' and 7' appear in the formula for the retarded density at the point c if $U_c>0$ and $M_4>M_1$. Alternatively, the values at the nodes 3', 6' and 9' appear when $U_a<0$ and $M_6>M_1$. In both these cases the central node 5 will be said to belong to the supersonic region. For the boundary cells the computational molecule is truncated by excluding the nodes 1', 1, 2, 3 and 3'.

The boundary conditions along the cut and the outer circle take the form:

$$\Phi(m,\eta)-\Phi(0,\eta)=\Gamma, \tag{2}$$

and

$$\Phi(\xi,n)=\lambda_\infty\bar{x} + \frac{\Gamma}{2\pi} \arctan\left((1-M_\infty^2)^{1/2}\bar{x}/\bar{y}\right), \tag{3}$$

respectively, where Γ: the velocity circulation, λ_∞ and M_∞: the free-stream values of λ and M, and $\bar{x}$, $\bar{y}$: the coordinates of the point x, y in a rotated Cartesian coordinate system with the $\bar{x}$-axis pointing in the direction of the free-stream velocity.

To determine the circulation Γ we additionally require that the pressure, and thus also the tangential velocity, should tend to the equal limits when the trailing edge is approached along the upper and the lower sides of the airfoil (the Kutta condition). Introducing the arc-length coordinate $s(\xi)$, measured along the airfoil surface in the counter-clockwise direction from the trailing edge, we can write

$$-\Phi_s(1/2,0) \simeq \Phi_s(m-1/2,\ 0),$$

or in the discrete form:

$$\Phi(0,0)-\Phi(1,0) = q(\Phi(m,0)-\Phi(m-1,0)), \tag{4}$$

where $q = s(1)/(s(m)-s(m-1))$ is the ratio of the mesh distances on both sides of the trailing edge. For simplicity the grid will always be constructed in such a way that $q = 1$.

Using (4) and (2) for $\eta = 0$ we get

$$\Gamma = \Phi(m-1,0)-\Phi(1,0). \tag{5}$$

3.2 Solution refinement on a succession of grids

Let us define the sequence of grids

$$G_i=\{(\xi,\eta):\xi=0,1,2,\ldots,m;\ \eta=0,\Delta\eta_i,2\Delta\eta_i,\ldots,n\},\ i=1,2,\ldots,k, \tag{6}$$

where $\Delta\eta_k=1$ and $\Delta\eta_{i-1}=2\Delta\eta_i$ for $i=k, k-1,\ldots,2$. The grid G_{i-1} is obtained by removing the even lines $\eta=(j-1)\Delta\eta_i$, $j=2,4,\ldots$, of the grid G_i. The horizontal lines η = const always contain the same number of equidistant nodes.

Let $\Phi^{(i)}$ denote the solution of the above presented discrete problem on a grid G_i, $i<k$. An iterative solution procedure on a finer grid G_{i+1} can be started with the initial approximation defined by

$$\Phi_o^{(i+1)}(\xi,\eta)=\Phi^{(1)}(\xi,\eta) \text{ for } \eta=0,2\Delta\eta_{i+1}, 4\Delta\eta_{i+1},\ldots,n, \quad (7a)$$

$$\Phi_o^{(i+1)}(\xi,\eta)=(\Phi^{(i)}(\xi,\eta-\Delta\eta_{i+1})+\Phi^{(i)}(\xi,\eta+\Delta\eta_{i+1}))/2 \quad (7b)$$

for $\eta = \Delta\eta_{i+1}, 3\Delta\eta_{i+1},\ldots,n-\Delta\eta_{i+1}$.

The choice of the grid sequence (6) can now be justified by the predicted properties of the solutions $\Phi^{(i)}$, $i\leq k$. For transonic flows the solutions exhibit shock waves, which are nearly orthogonal to the airfoil surface and thus nearly parallel to the lines ξ=const. The solutions are smooth along these lines and, consequently, the prolongations of the form (7 b) are not likely to introduce large initial errors into $\Phi_o^{(i+1)}$.

Nevertheless, the averages (7 b) seem to be less trustworthy than the values of the converged solution $\Phi^{(i)}$ in (7 a). Therefore, these averages will only be used as an initial approximation, required by the second stage of the prolongation procedure.

The second stage consists in solving the mass-conservation equations (1 a) for the control-cells centered at the grid-points of the even lines $\eta=\Delta\eta_{i+1}, 3\Delta\eta_{i+1},\ldots$, with the fixed values (7a) of the potential along the odd lines. The independent nonlinear subsystems for the values of the potential along the even lines are supplemented by the boundary conditions (2) with the value of Γ such as that for $\Phi^{(i)}$, and solved by the Newton method with the starting approximation (7 b) (one iteration of the Newton method was always sufficient). Such a "natural" prolongation procedure takes full account of the definition of the computational problem.

The subsequent iterative process consists in repeating the second stage of the prolongation procedure alternatively for odd and even lines ("zebra"-line-Newton iteration).

Starting each iteration with the line η=0 we solve the system (1b) with the Kutta boundary condition (4). The given values $\Phi_n^{(i+1)}(\xi,\Delta\eta_{i+1})$ of the previous approximation along the second line are substituted into (1b). As for the other lines, the solution $\Phi_{n+1}^{(i+1)}(\xi,0)$ immediately overwrites the previous values $\Phi_n^{(i+1)}(\xi,0)$. The external distribution (3) of the potential is then corrected by substituting the updated value Γ_{n+1}, given by

(5). The fixed value Γ_{n+1} is used in the boundary conditions (2) for the inner lines $\eta=2\Delta\eta_{i+1}$, $4\Delta\eta_{i+1}$, ..., and then $\eta=\Delta\eta_{i+1}$, $3\Delta\eta_{i+1}$,..., until the next visit to the line $\eta=0$. The best efficiency and stability was achieved with the quasi-Newton method, reduced to a single Newton iteration per line.

If $i<k-1$ then the converged result $\Phi^{(i+1)}$ of this iterative procedure is prolongated to the grid G_{i+2} and the procedure is applied again. The overall calculation starts with $\Phi_0^{(1)}=\lambda_\infty\cdot\bar{x}$, corresponding to the case of the uniform flow, and ends with the sufficiently converged result $\Phi^{(k)}$.

Except for several m-element double precision arrays, needed for the solution of the Newton-linearized systems, the present method uses only one $m \times (n+1)$-element single-precision array, necessary for the allocation of the final result $\Phi^{(k)}$ and all the intermediate distributions of the potential.

3.3 Newton iterations

For a fixed line η=const the Newton-linearized systems (1a) or (1b) can be written in the form:

$$a_i x_{i-2}+b_i x_{i-1}+c_i x_i+d_i x_{i+1}+e_i x_{i+2}=f_i, \quad i=1,2,\dots,m, \tag{8}$$

where x_i denotes the current Newton correction at $\xi=i$, and a_i, $b_i,\dots,f_i$ are the appropriate elements of the Jacobian matrix and the residue.

For the line $\eta=0$, the correction satisfies the Kutta condition (4) with $q=1$, which takes the form:

$$x_0 - x_1 = x_m - x_{m-1}. \tag{9}$$

For the other lines, the Kutta condition is replaced by (2) which gives

$$x_m-x_0= \gamma, \tag{10}$$

where γ is the given difference $\Gamma_{n+1}-\Gamma_n$ of the last approximants for Γ.

For all the lines we must also satisfy the condition

$$x_{m+1} - x_m = x_1 - x_0, \tag{11}$$

which follows from the periodicity of the velocity field with respect to the ξ-coordinate.

It is assumed that the flow is subsonic near the gridline $\xi=0$, which issues from the trailing edge of the airfoil. As a consequence we have

$$a_1=e_1=a_2=e_{m-1}=a_m=e_m=0. \tag{12}$$

For the line $\eta=0$ the conditions (9) and (11) give

(13) $$x_{m+1}=x_{m-1}, \tag{13}$$

$$x_{m-1} = x_m - x_o + x_1 . \tag{14}$$

Using (12) and (13) we can put the last equation of the system (8) into the form

$$(b_m + d_m)x_{m-1} + c_m x_m = f_m, \tag{15}$$

which, when combined with (14), gives

$$x_o = x_1 + \alpha x_m + \beta, \tag{16}$$

where

$$\alpha = (b_m + c_m + d_m)/(b_m + d_m), \qquad \beta = -f_m/(b_m + d_m).$$

Substituting (16) into the first equation of the system (8) we obtain

$$b_1\alpha x_m + (b_1 + c_1)x_1 + d_1x_2 = f_1 - b_1\beta. \tag{17}$$

Now, replacing the first and the last equation in (8) with (17) and (15), respectively, we obtain a complete system of periodic structure for $x_1, x_2, \ldots, x_m$.

Using (10) and (11) we obtain an analogous system for the other lines, with the first and the last equation of the respective forms:

$$b_1x_m + c_1x_1 + d_1x_2 = f_1 + b_1\gamma,$$

$$b_mx_{m-1} + c_mx_m + d_mx_1 = f_m - d_m\gamma.$$

We shall now present an exact method for solving such systems. Let the system (8), corrected for periodicity, be represented in the form $Lx=f$, with

$$L = aS_\xi^-S_\xi^- + bS_\xi^- + cI + dS_\xi^+ + eS_\xi^+S_\xi^+, \tag{18}$$

where a, b, ..., e denote the appropriate periodic grid functions on a line, and S_ξ^+, S_ξ^- are the shift operators defined by:

$$(S_\xi^+x)_i = x_{i+1}, \quad (S_\xi^-x)_i = x_{i-1} \quad \text{for } i = 0, \pm1, \pm2, \ldots$$

The operator L will be decomposed into the product

$$L = (\tilde{a}S_\xi^- + I + \tilde{e}S_\xi^+)(\tilde{b}S_\xi^- + \tilde{c}I + \tilde{d}S_\xi^+) . \tag{19}$$

Comparing (18) and (19) we get

$$\tilde{a}(S_\xi^- \tilde{b}) = a,$$

$$\tilde{e}(S_\xi^+ \tilde{d}) = e,$$

$$\tilde{a}(S_\xi^- \tilde{c}) + \tilde{b} = b,$$

$$\tilde{a}(S_\xi^- \tilde{d}) + \tilde{c} + \tilde{e}(S_\xi^+ \tilde{b}) = c,$$

$$\tilde{d} + \tilde{e}(S_\xi^+ \tilde{c}) = d.$$

We shall consider the three possible cases:

(i) the point (i,η) belongs to the subsonic region; then

$a_i = e_i = 0$ and the above formulas give

$$\tilde{a}_i = \tilde{e}_i = 0;$$

$$\tilde{b}_i = b_i,\ \tilde{c}_i = c_i,\ \tilde{d}_i = d_i;$$

(ii) the point (i,η) belongs to the supersonic region on the upper surface of the airfoil, where $U<0$; then $a_i=0$, $e_i \neq 0$ and, consequently,

$$\tilde{a}_i = 0,$$

$$\tilde{e}_i = e_i/\tilde{d}_{i+1},$$

$$\tilde{b}_i = b_i, \tag{20a}$$

$$\tilde{c}_i = c_i - \tilde{e}_i \tilde{b}_{i+1}, \tag{20b}$$

$$\tilde{d}_i = d_i - \tilde{e}_i \tilde{c}_{i+1}; \tag{20c}$$

(iii) the point (i,η) belongs to the supersonic region on the lower surface, where $U>0$; then $a_i \neq 0$, $e_i=0$ and

$$\tilde{a}_i = a_i/\tilde{b}_{i-1}, \tag{21a}$$

$$\tilde{e}_i = 0,$$

$$\tilde{b}_i = b_i - \tilde{a}_i \tilde{c}_{i-1}, \tag{21b}$$

$$\tilde{c}_i = c_i - \tilde{a}_i \tilde{d}_{i-1}, \tag{21c}$$

$$\tilde{d}_i = d_i.$$

The system $Lx = f$ may now be rewritten in the form

$$(\tilde{b}S_\xi^- + \tilde{c}I + \tilde{d}S_\xi^+)x = \tilde{f}, \tag{22}$$

where

$\tilde{f}i = f_i$ in the case (i),

$$\tilde{f}_i = f_i - \tilde{e}_i \tilde{f}_{i+1} \qquad \text{in the case (ii), and} \qquad (23a)$$

$$\tilde{f}_i = f_i - \tilde{a}_i \tilde{f}_{i-1} \qquad \text{in the case (iii).} \qquad (23b)$$

Let us set $\tilde{b}_i = b_i$, $\tilde{c}_i = c_i$, $\tilde{d}_i = d_i$ and $\tilde{f}_i = f_i$ for all $i = 1, 2, \ldots, m$. Proceeding in the direction of the increasing index $i = 1, 2, \ldots, m$ we introduce the corrections defined by (21a, b, c) and (23b), whenever $a_i \neq 0$. The analogous procedure in the reverse direction, using (20a, b, c) and (23a) for $e_i \neq 0$, completes the calculation of the coefficients and the right-hand sides in (22).

The corresponding equations of the systems (8) and (22) differ only at the points (i,η) belonging to the supersonic flow regions. The corrections which must be introduced at a supersonic point cost only 7 arithmetic operations.

The exact solution of the three-diagonal matrix system (22) of periodic structure is obtained by the method of the recursive cyclic reduction [4], specialized to the case of m equal to a power of 2. The method requires 17 arithmetic operations per equation. Hence, the solution of (8) costs much less than the calculation of the residues f_i.

In an economic computer code the coefficients a_i, b_i, ..., e_i are obtained cheaply as a by-product of the residue calculation. The amount of the computational work equivalent to a complete iteration of the line-Newton method on the finest grid will be called a "work unit".

3.4 Results of the calculations

Among other flow cases, the calculations were carried out for the transonic flow around the NACA 0012 airfoil with the free-stream Mach number $M_\infty=0.75$, at the angle of attack $\alpha_\infty=2^\circ$. The three grids were used, containing 6, 11 and 21 horizontal lines, respectively, and 64 vertical lines on the computational rectangle.

As a measure of the solution error we chose the absolute sum of the residues in the equations (1a,b). Such a measure of error, which is an approximation of the L_1-norm of the densities of the erroneous mass sources in the flow region, seems to be relatively independent of the grid resolution.

In our calculations, the two order of magnitude reduction of the initial error (on the coarsest grid) was obtained on a finest grid as a result of 72.75 work units (when repeated on the Amdahl 470 computer, the calculation required 34 CPU-seconds).

Fig. 2 represents the comparison of the present results for the pressure coefficient distribution on the airfoil surface with the results of the more accurate calculations of our earlier report [7] or the paper [8], performed on a finer grid containing 41 horizontal and 128 vertical lines. The present results for the lift coefficient: $c_L=0.5685$ are fairly close to the results: $c_L=0.5596$ of [8], $c_L=0.571$ of [6] and $c_L=0.5878$ of

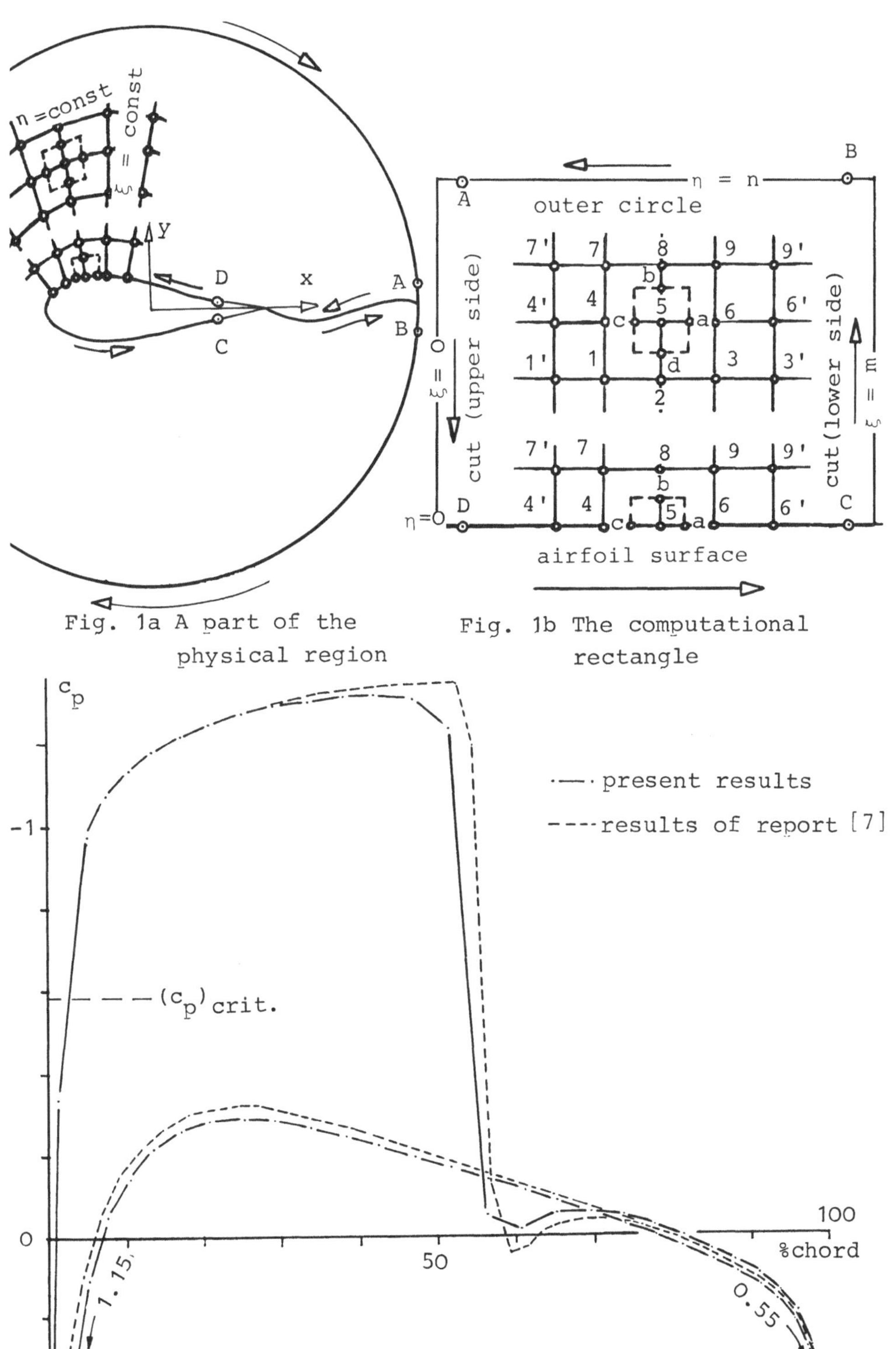

Fig. 1a A part of the physical region

Fig. 1b The computational rectangle

Fig. 2 The distribution of the pressure coefficient on the surface of the NACA 0012 airfoil; $M_\infty = 0.75$, $\alpha = 2^\circ$.

[5]. The value of the drag coefficient depends more critically on the grid resolution and geometry. The present result c_D=0.0214 is slightly larger than the results c_D=0.0163, c_D=0.0175 and c_D=0.0182 of [8], [6] and [5], respectively.

For the finer grid of the size 41 × 128 a very slow (practically stalling) convergence was observed during the final phase of the calculation (switching to the double precision accuracy in the whole program did not introduce any visible improvements). Such a behaviour is due to the low "numerical signal speed" along the lines ξ=const, characteristic of the present method. For the present "zebra"-line relaxation procedure a correction introduced into the outer boundary distribution (3) starts to be felt on the airfoil surface after n/2 complete iterations on the finest grid. Perhaps the line-Gauss-Seidel relaxation with the alternating sweep direction could be more efficient. It seems, however, that a substantial increase of the "numerical signal speed", while maintaining the very modest storage requirements, can only be achieved with the aid of the multigrid methods, such as those presented in [2].

4. THE COMPUTATIONAL METHOD FOR MULTI-ELEMENT AIRFOILS

4.1 General presentation

In the case of a multi-element airfoil the flow region is divided into the exterior of the "extended main contour" (bold line in Fig. 3) and the channel regions between the main airfoil and its satellites.

Fig. 3 A multi-element airfoil and the extended main contour.

The "zebra"-line-Newton method of Section 3.2 was generalized by including a new "line", corresponding to the totality of the grid-points in the channel regions. Consequently, the treatment of the even lines is started with a single iteration of the line-Newton method in the channels. The procedure for each channel is similar to that for the exterior region.

A curvilinear grid in a channel (Figs. 4,9) is generated

as the image of the regular grid on a rectangle $0 \le \xi_c \le m_c$, $0 \le \eta_c \le n_c$, under the transformation defined by the formula

$$\begin{aligned} x(\xi_c,\eta_c) = {} & x(\xi_c,0) + x(0,\eta_c) - x(0,0) + \eta_c(x(\xi_c,n_c) \\ & -x(\xi_c,0) + x(0,0) - x(0,n_c))/n_c \\ & + \xi_c(x(m_c,\eta_c) - x(m_c,0) - x(0,\eta_c) \\ & + x(0,0))/m_c + \xi_c\eta_c(x(m_c,0) - x(m_c,n_c) \\ & + x(0,n_c) - x(0,0))/(m_c n_c), \end{aligned} \tag{24}$$

and the analogous formula for $y(\xi_c,\eta_c)$. The chosen functions $x(\xi_c,0)$, $y(\xi_c,0)$, $x(0,\eta_c)$,..., appearing on the right-hand sides of these formulas represent the correspondence of the boundaries of the computational rectangle and the channel region.

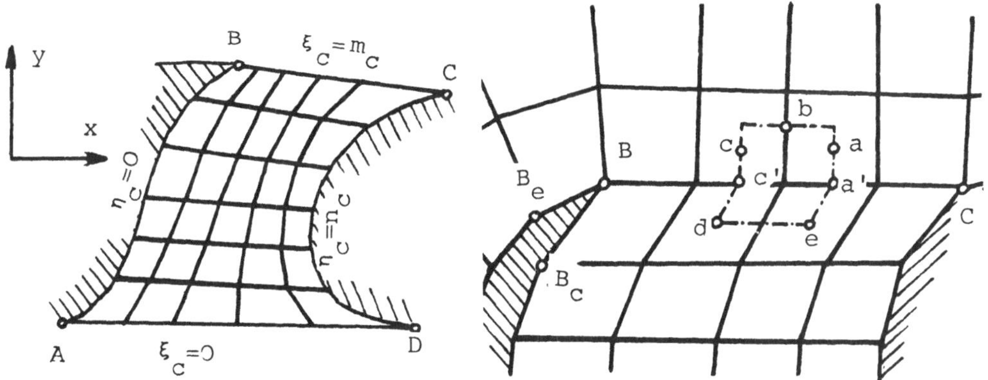

Fig. 4 The curvilinear grid in a channel.

Fig. 5 A control-cell at the boundary between the channel and the exterior region.

An iteration in a channel region is started with the line $\eta_c = 0$. Assuming that the distribution of the potential along the next line $\eta_c = 1$ is given, the equations (1b) for the surface values of the potential are linearized by the Newton method. It is assumed that the velocity vector always points in the direction from A to B, which implies that $e_i = 0$, $i = 1,2,\ldots,m_c - 1$, in (8). The system (8) is led to the three-diagonal matrix form by the method of Section 3.3. It should be mentioned that the present approach requires the additional assumption that the flow is subsonic at $\xi_c = 1$, which may not always be true.

Let Φ_c and Φ_e denote the distributions of the potential inside a channel and the exterior region, respectively. If we assume that Φ_c and Φ_e agree along the line AD then a constant increment $\Phi_c - \Phi_e = \Gamma_c$ must be imposed along the line BC. The value Γ_c follows from the Kutta condition at B, which in analogy with (4) can be written as:

$$(\Phi_e(B)-\Phi_e(B_e))=q(\Phi_c(B)-\Phi_c(B_c)), \tag{25a}$$

where q is the ratio of the mesh distances on both sides of B (see Fig. 5), and

$$\Phi_c(B) = \Phi_e(B)+ \Gamma_c. \tag{25b}$$

The three-diagonal matrix system is solved using the boundary condition

$$\Phi_c(A) = \Phi_e(A)$$

and (25a), with the values $\Phi_e(A)$, $\Phi_e(B)$ and $\Phi_e(B_e)$ given by the previous iteration in the exterior region. Now, Γ_c follows from (25b).

Usually, one or two iterations of the Newton method were sufficient to obtain a practically converged distribution $\Phi_c(\xi_c,0)$ and the value of Γ_c. The subsequent procedure for the lines $\eta_c=1,2,...,n_c$ is analogous, except that the Kutta condition (25a) is now replaced by (25b) with the given value of Γ_c. In the transonic flow cases the Newton corrections were underrelaxed for stability, with the underrelaxation coefficient ranging from 0.5 to 0.7.

When all the lines in the channels have been thus treated, the calculation is continued for the lines η=const in the exterior region, as in the case of a single airfoil. The line $\eta=0$ now corresponds to the extended main contour, embracing the airfoils and the channel regions. The mass conservation equations (1b) for the control-cells centered at the grid-points of the segments such as AD and BC in Fig. 4 must be corrected by including the mass fluxes across these segments. For a cell, represented in Fig. 5, the corrected equation can be written as

$$\frac{1}{2}(\tilde{\rho}U/I)_a - \frac{1}{2}(\tilde{\rho}U/I)_c + (\rho V/I)_b = -Q, \tag{26}$$

where Q denotes the given mass flux across the segments dc', ea' and de, calculated using the previously determined distribution Φ_c of the potential inside the channel.

The overall calculation is organized in the form of the solution refinement procedure, using a succession of grids in the exterior region and the fixed grids in the channels.

The calculation on the coarsest grid is started with $\Phi_e=U_\infty \bar{x}$ and $\Phi_c=0$ in all the channels.

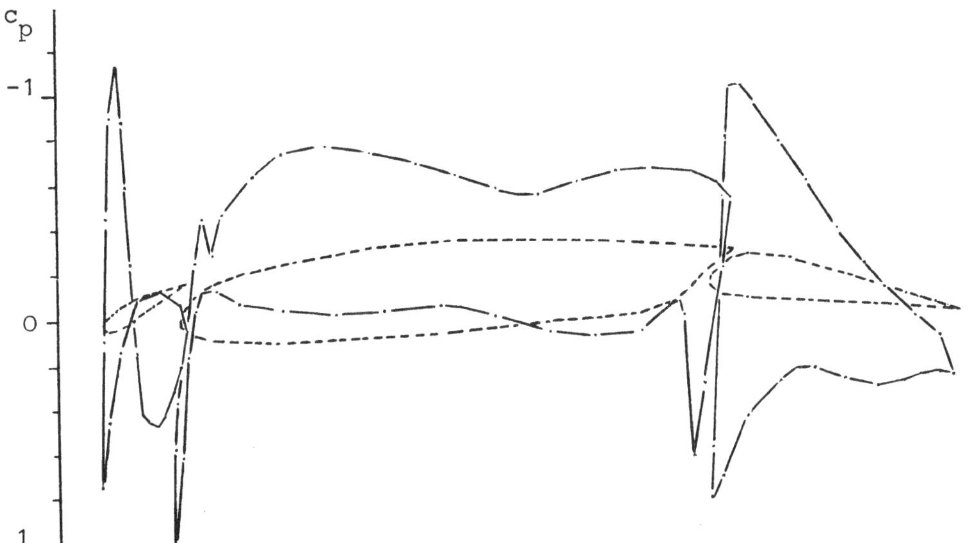

Fig. 6 The c_p-distribution on the surface of the three element airfoil (data of the Warsaw Institute of the Aircraft); $M_\infty = 0.4$, $\alpha = 0^o$.

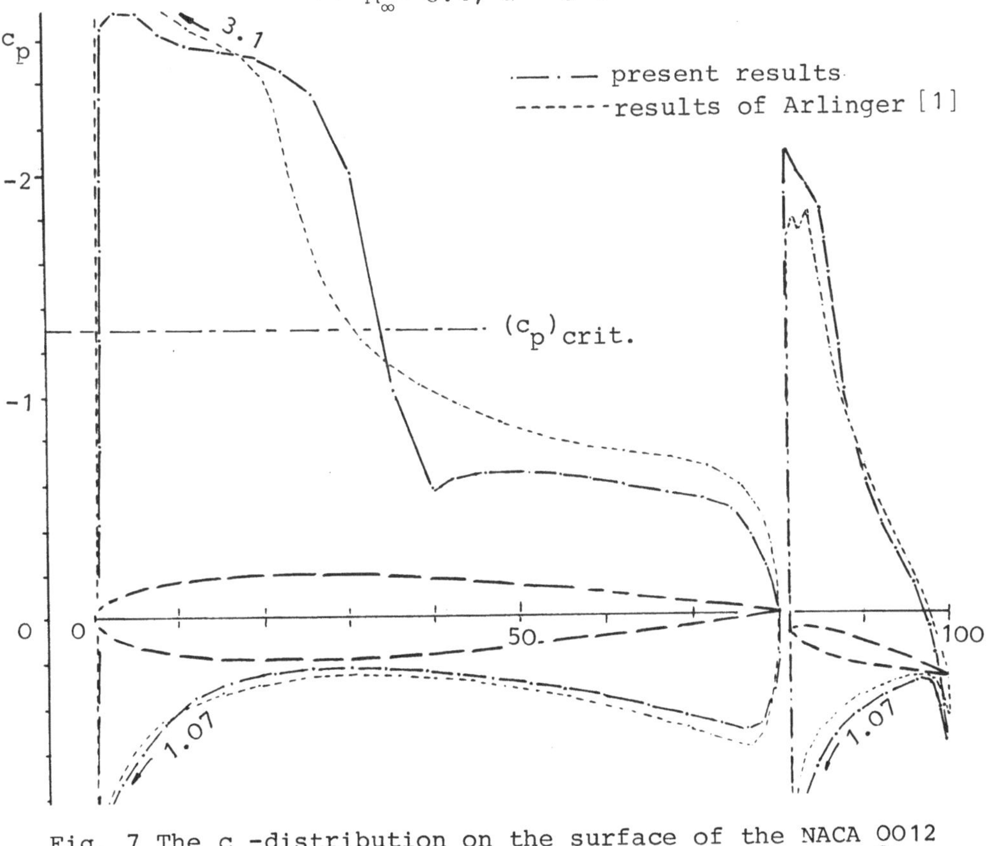

c_p

present results

results of Arlinger [1]

3.1

-2

$(c_p)_{crit.}$

-1

0 0 50 100

1.07

1.07

Fig. 7 The c_p-distribution on the surface of the NACA 0012 airfoil with the NACA 0012 flap; $M_\infty = 0.6$, $\alpha = 2^o$.

4.2 Results of the calculations

The present method was applied in a variety of the subsonic and the transonic flow cases, for the two and the three-element airfoils.

The three grids were used in the exterior region, containing 5,9 and 17 lines η=const, and 64 lines ξ=const. An example of a grid in a channel is shown in Fig. 9.

The results for the c_p-distributions on the surface for a subsonic and a transonic flow case are shown in Figs. 6 and 7.

Fig. 7 represents the comparison of the present results with the results of the paper [1], obtained with the aid of the conformal mapping techniques. A more pronounced shock wave on the upper surface of the main airfoil can be seen in our results.

For stability, the present method required a certain skill in choosing the shape of the channel in such cases as that represented in Fig. 9. As a rule, a few hundred work units were needed to obtain a sufficiently converged solution.

5. GRID GENERATION BY THE SEMI-ANALYTIC FORMULAS

5.1 General presentation of the method

For the grid generation in the exterior regions we used the mapping functions of the form

$$x(\xi,\eta)=a_1(\xi) + a_2(\xi)(\cosh(c_1\eta)-1)+ a_3(\xi)(\exp(-c_1\eta)-1), \tag{27a}$$

$$y(\xi,\eta)=a_4(\xi) + a_5(\xi)(\cosh(c_1\eta)-1)+a_6(\xi)(\exp(-c_1\eta)-1). \tag{27b}$$

As a consequence of the formulas

$$x(\xi,0)=a_1(\xi), \; y(\xi,0)=a_4(\xi), \tag{28a}$$

$$x_\eta(\xi,0)= -c_1a_3(\xi), \; y_\eta(\xi,0) = -c_1a_6(\xi), \tag{28b}$$

the properties of the transformation (27a,b) near the line η=0 are determined by the functions $a_1(\xi)$, $a_4(\xi)$, $a_3(\xi)$ and $a_6(\xi)$.

For the appropriate choice of $a_1(\xi)$ and $a_4(\xi)$ this line can be mapped on an arbitrary contour, such as the airfoil surface, and any desired distribution of the boundary grid-points can be obtained. Furthermore, if we choose

$$a_3(\xi) = -\frac{a_7(\xi)}{c_1}\frac{da_4}{d\xi}, \tag{29a}$$

$$a_6(\xi) = \frac{a_7(\xi)}{c_1}\frac{da_1}{d\xi}, \tag{29b}$$

then by (28a,b) it will follow that

$$x_\xi x_\eta + y_\xi y_\eta = 0 \qquad \text{for } \eta = 0,$$

which means that the resulting grid is orthogonal at the points of the airfoil surface. In practice, the grid remains nearly orthogonal in the whole region (see Fig. 8).

Using (28a,b) and (29a,b) we obtain

$$a_7(\xi) = ((x_\eta^2+y_\eta^2)/(x_\xi^2 + y_\xi^2))^{1/2} \simeq h/w \text{ for } \eta = 0,$$

where h and w denote the mesh sizes in the appropriate directions. Hence, choosing $a_7(\xi)$ we may easily control the mesh form near the airfoil.

The remaining two functions $a_2(\xi)$ and $a_5(\xi)$ can be chosen so as to enforce the correspondence of the line $\eta=n$ and the outer boundary of the region. For large η we have

$$x(\xi,\eta) \simeq a_2(\xi) \exp(c_1\eta)/2, \tag{30a}$$

$$y(\xi,\eta) \simeq a_5(\xi) \exp(c_1\eta)/2. \tag{30b}$$

Hence, if we put

$$a_2(\xi) = 2c_2 \cos(2\pi\xi/m)/c_1, \tag{31a}$$

$$a_5(\xi) = 2c_2 \sin(2\pi\xi/m)/c_1, \tag{31b}$$

then the line $\eta=n$ will be mapped on a near circle of the approximate radius

$$R = c_2 \exp(c_1 n)/c_1. \tag{32}$$

If η is large then, as it follows from (30a, b) and (31a,b), we have

$$c_1 \simeq \frac{2\pi}{m}((x_\eta^2 + y_\eta^2)/(x_\xi^2+y_\xi^2))^{1/2} \simeq \frac{2\pi}{m}\frac{h}{w} .$$

Consequently, c_1 determines the mesh form far from the airfoil. The remaining constant c_2 follows from the requirement (32) with a given value of R.

For cambered airfoils the intersections of the grid lines orthogonal to the airfoil surface could result inside the region. To avoid this we treat the functions (27a,b) with an averaging operation. For all the grid points outside the airfoil surface we use the corrected formulas (27a,b) with the functions $a_1(\xi)$, $a_2(\xi)$,..., $a_6(\xi)$ replaced by

$$\bar{a}_i(\xi) = a_i(\xi) + a_8(\xi)((a_i(\xi-1)+a_i(\xi+1))/2-a_i(\xi)), \tag{33}$$
$$i=1,2,...,6,$$

where $a_8(\xi)$ is a smooth function which may vanish outside certain intervals. Smooth and nearly orthogonal grids are obtained if the operation (33) is repeated several times with small values of $a_8(\xi)$.

If the functions $a_i(\xi)$ or $\overline{a_i}(\xi)$, i=1,2,...,6 and the accompanying factors in (27a,b) have been determined in advance, then the subsequent calculation of a pair $x(\xi,\eta)$, $y(\xi,\eta)$ requires only 8 arithmetic operations. Therefore, instead of storing the functions x and y, we preferred to recompute their values in the programs presented in Sections 3 and 4.

5.2 Examples of grids

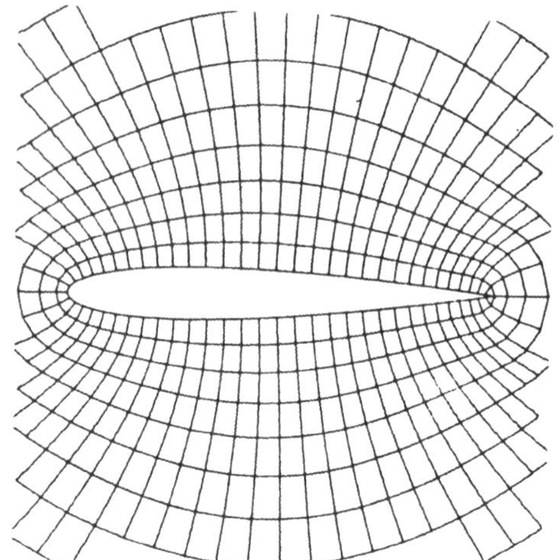

Fig. 8 A part of the grid around the NACA 0012 airfoil.

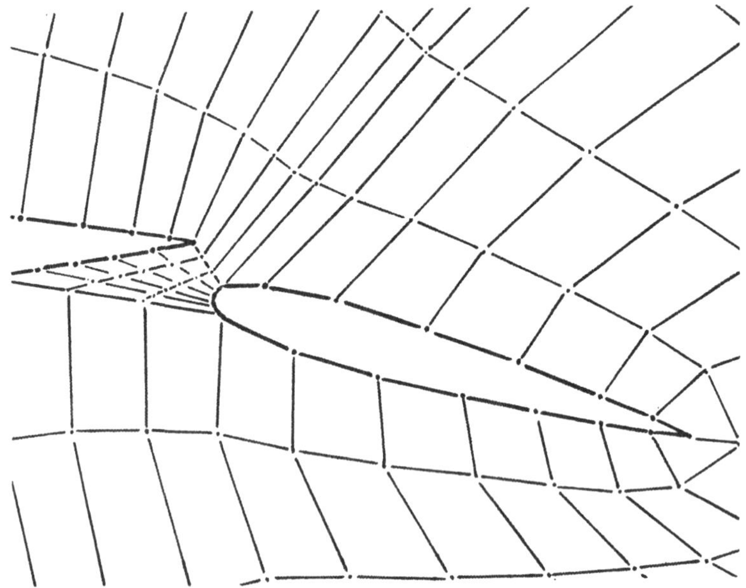

Fig. 9 A part of the composite grid near the flap of a multi-element airfoil.

Fig. 8 represents a part of the grid generated around the NACA 0012 airfoil with the aid of the formulas (27a,b).The present method was also used for generating the grids in the exterior of the multi-element airfoils, i.e., outside the union of their components and the channels between them. Fig. 9 represents a part of the grid near the flap of a multi-element airfoil.

Local grid averaging (33) was applied to avoid the grid intersections or irregularities near the inlet and the outlet of the channel region. The grid in the channel was obtained with the aid of the formula (24) for the x-coordinate and the analogous formula for the y-coordinate. The resulting composite grid was used for the transonic flow calculations by the methods of Section 4, in the flow case corresponding to Fig. 7.

Acknowledgement. Thanks are due to Professor Pieter Wesseling for his kind help during the visit of the present author at the Department of Mathematics and Informatics of Delft University of Technology. This research was partly supported by the Netherlands Organization for the Advancement of Pure Research (ZWO).

REFERENCES

[1] ARLINGER, B.G.: "Analysis of two-element high lift systems in transonic flow", ICAS Paper 76-13 (1976).

[2] BRANDT, A.: "Multi-level adaptive solutions to boundary value problems", Math. Comp., 31 (1977) pp. 333-390.

[3] HACKBUSCH, W.: "Multi-grid convergence theory", in: Multigrid Methods, Proceedings, Cologne 1981, Hackbusch, W., Trottenberg, U., eds., Lecture Notes in Mathematics 960, Springer-Verlag, Berlin 1982, pp. 177-219.

[4] HOCKNEY, R.W.: "A fast direct solution of Poisson's equation using Fourier analysis", Journal of the Association for Computing Machinery, 12 (1965) pp. 95-113.

[5] HOLST, T.L.: "An implicit algorithm for the conversative, transonic full potential equation using an arbitrary mesh", AIAA Paper 78-1113 (1978).

[6] JAMESON, A.: "Acceleration of transonic potential flow calculations on arbitrary meshes by the multiple grid method", AIAA Paper 79-1458 CP (1979).

[7] NOWAK, Z.: "A quasi-Newton multigrid method for determining the transonic lifting flows around airfoils", Delft University of Technology Report 84-11 (1984).

[8] NOWAK, Z., WESSELING, P.: "Multigrid acceleration of an iterative method with application to transonic potential flow", in: Computing Methods in Applied Sciences and Engineering, VI, Proceedings, Versailles 1983, Glowinski, R., Lions, J.L., eds., North Holland, Amsterdam 1984, pp. 199-217.

BASIC SMOOTHING PROCEDURES FOR THE MULTIGRID TREATMENT OF ELLIPTIC 3D-OPERATORS

Clemens-August Thole
Ulrich Trottenberg
GMD, Institut für Methodische Grundlagen(F1)
Schloß Birlinghoven, D-5205 Sankt Augustin 1, Germany

SUMMARY

From 2D-multigrid it is a fundamental insight that pointwise relaxation combined with standard coarsening gives no reasonable smoothing effect as soon as the given elliptic operator has a considerably *anisotropic* behavior. Corresponding questions and practical consequences are discussed in this paper for the 3D-case: In the general anisotropic 3D-case, even line relaxation is not sufficient if standard coarsening is maintained. Instead, *"plane relaxation"* is necessary in certain cases. If plane relaxation is applied correctly and performed by use of appropriate 2D-multigrid, the resulting 3D-multigrid method has an asymptotic complexity of O(N) (where N = number of 3D-grid points) and is – indeed – highly efficient already for moderate values of N. In contrast to the common opinion, plane relaxation turns out to be a simple and general smoothing concept for standard elliptic 3D-problems.

1. INTRODUCTION: THE 2D-CASE

The multigrid (MG) principle is based on smoothing and approximation on coarser grids and yields "optimal" solution algorithms [3]. In order to obtain not only asymptotically optimal methods, but really efficient ones, it is necessary to tune the smoothing and coarse-grid approximation processes in a suitable problem-dependent way. A frequently arising, especially simple and fundamental situation in which this necessity is very well-known is the case of a 2D scalar anisotropic operator. Let us consider the model operator

$$a u_{xx} + b u_{yy}$$

with

$$0 < a \ll b \quad \text{or} \quad 0 < b \ll a.$$

For such an operator, the smoothing properties of pointwise relaxation schemes are bad, if standard coarsening is used. (This has been proved by model problem analysis [4], local Fourier analysis [4], [1], and is well-known in the MG practice.) The reason for this is that pointwise relaxation has, in this case, a smoothing effect only with respect to the "dominant direction" of the operator, i.e. the

$$\text{y-direction for } \; 0 < a \ll b,$$

$$\text{x-direction for } \; 0 < b \ll a.$$

(Brandt usually denotes this dominance as "strong coupling".)

There are two alternative ways in which this difficulty can be mastered. Firstly, instead of standard coarsening, *semi-coarsening* can be used to obtain a resonable smoothing effect. More specifically: The pointwise relaxation is maintained, but the coarse grid is defined by doubling the meshsize only with respect to the dominant direction. That this gives a reasonable smoothing effect, can be seen from the following formal consideration (also see the formal smoothing analysis as described e.g. in [4], Chapter 7): Since there is no coarsening performed with respect to the non-dominant direction, there are no high-frequency components defined and - therefore - no smoothing is required, *with respect to this direction.*

The second possibitity is to maintain the standard coarsening strategy but to use *linewise* relaxation instead of pointwise relaxation. Choosing the lines in the dominant direction, all unknowns of any such gridline are resolved simultaneously in this approach and one obtains the desired smoothing effect also with respect to the non-dominant direction. Practically, the occuring linear systems that have to be solved (in one relaxation step for each gridline) are tridiagonal or, more generally, bandstructured with constant bandwidth. In other words: They correspond to 1D-boundary value problems. Since all these systems can be solved (by bandsolvers) in a number of operations that is proportional to the number of unknowns, these direct solution processes don't disturb the principal asymptotic optimality of multigrid.

Users often prefer the line relaxation approch with standard coarsening (rather than the semi-coarsening approach with pointwise relaxation), as this approach is regarded as somewhat simpler and more robust. Robustness is the essential objective, in particular, in using *alternating line relaxation.* This applies to the situation that - for variable coefficients a,b - both relations $0 < a \ll b$ and $0 < b \ll a$ are true (in different parts of the considered domain). (Clearly, if in such cases one would try to use semi-coarsening in different directions for certain parts of the domain, one would obtain a very complicated grid structure. Such data-dependent grid structures are natural within the AMG approch [5], but are usually avoided in "classical" geometric MG.)

Altogether, one may conclude that, with respect to the MG-treatment of anisotropic operators, the 2D-case is settled and well understood.

2. THE 3D-SITUATION, TWO-GRID ANALYSIS

The principal phenomena in the 3D-case are similar to those in the 2D case. However, the number of different kinds of anisotropic behavior to be distinguished is larger. Only very few systematical investigations on how to treat such anisotropic 3D-problems have been carried out so far. Here we want to give quantitative results (two-grid convergence analysis) for the model operator

$$au_{xx} + bu_{yy} + cu_{zz}. \tag{M}$$

We distinguish the four situations:

(1) $0 < a \sim b \sim c$ (a,b,c same order of magnitude),
(2) $0 < a \sim b \ll c$ (a,b same order of magnitude, c essentially larger),
(3) $0 < a \ll b \sim c$ (b,c same order of magnitude, a essentially smaller),
(4) $0 < a \ll b \ll c$ (a essentially smaller than b, c essentially larger than b).

For constant coefficients, all other cases can be reduced to these four situations. From the general considerations of the previous section, we may conclude the following statements heuristically:

(1) In this case, standard coarsening and (suitable) pointwise relaxation give good smoothing.
(2) (Suitable) pointwise relaxation gives good smoothing only in z-direction in this case. To achieve good overall smoothing, again two approches are feasible: standard coarsening combined with (suitable) z-line relaxation or (suitable) pointwise relaxation combined with z-semi-coarsening (i.e. coarsening only in z-direction).
(3) (Suitable) pointwise relaxation gives good smoothing with respect to y and z, but not with respect to x in this case. In contradistinction to the previous case, here even any line relaxation is unsatisfactory if used in connection with standard coarsening. Instead, standard coarsening has to be combined with "(y,z)-plane relaxation". This means that, for each (y,z)-plane within the 3D-grid, all unknowns of the corresponding (y,z)-plane have to be solved for simultaneously. In other words, in each plane relaxation step, (n-1) 2D-problems have to be solved where n-1 is the number of (y,z)-planes in the grid. An alternative way of smoothing is to combine (suitable) pointwise relaxation with (y,z)-semi-coarsening.
(4) (Suitable) pointwise relaxation gives good smoothing only in z-direction, so that it may be combined only with z-semi-coarsening. Standard coarsening requires (y,z)-plane relaxation. Furthermore, (y,z)-semi-coarsening requires z-line relaxation.

The above considerations are made precise by the quantitative results in the following Table 1. We consider the 3D model operator with typical anisotropic constant coefficients and corresponding representative MG-components. Table 1 gives the convergence factors (more precisely: their h-independent upper limits)

$$\rho^* = \sup_h \rho(M_h^H)$$

for a variety of *two-grid* methods. Using the notation from [4], the two-grid methods are characterized by the iteration operator

$$M_h^H = S_h^{\nu_2}[I_h - I_H^h L_H^{-1} I_h^H L_h] S_h^{\nu_1} \quad (\rho(M_h^H) = \text{ spectal radius of } M_h^H).$$

Result 1. *We consider the Dirichlet problem*

$$Lu = f^\Omega \quad (\Omega), \qquad u = f^\Gamma \quad (\Gamma)$$

in the unit cube $\Omega = (0,1)^3$, $\Gamma = \partial\Omega$, *where L is the model operator (M) with constant coefficients* $a, b, c > 0$. *We assume that the problem is discretized on a* Ω*-matching cubic grid* Ω_h *of meshsize* $h = \frac{1}{n}$ $(n = 2^p, p = 2, 3, \ldots)$, *by use of the ordinary 7-point-approximation* L_h *of order* h^2. *As for the coarse grid* Ω_H, *we consider three cases of coarsening in Table 1, namely*

standard-, (y,z)-semi- and z-semi-coarsening.

In all cases, L_H *is assumed to be the ordinary 7-point approximation of L corresponding to* Ω_H. *The grid transfers* I_h^H *and* I_H^h *are made by full weighting and linear interpolation, respectively.*

As for relaxation (operator S_h), we consider the following possibilities:

point:	3D-red-black pointwise relaxation
z-line:	z-linewise relaxation (in a 2D-red-black order of lines)
(y,z)-plane:	(y,z)-plane relaxation (in a zebra order of planes).

Apart from these, in part (3) of Table 1 we apply

alt y/z-line: alternating y- and z-linewise relaxation (both in a 2D-red-black order of lines),

just to show that in this case even alternating linewise relaxation is not useful.

The parameter $\nu = (\nu_1, \nu_2)$ (numbers of relaxation steps before and after the coarse-grid correction, resp.) is fixed: $\nu_1 = \nu_2 = 1$, in Table 1.

TABLE 1: 3D-two-grid convergence factors ρ^* reflecting the four cases (1) – (4) distinguished above

	a	b	c	coarsening	relaxation $\nu = (1,1)$	ρ^*	comment
(1)	1	1	1	**standard**	**point**	0.198	correct
(2)	1	1	100	standard	point	0.961	wrong
				standard	**z-line**	0.074	correct
				(y,z)-semi	point	0.961	wrong
				z-semi	*point*	0.017	correct
(3)	1	100	100	standard	point	0.980	wrong
				standard	alt y/z-line	0.938	wrong
				standard	**(y,z)-plane**	0.052	correct
				(y,z)-semi	*point*	0.074	correct
(4)	1	100	10000	standard	point	1.000	wrong
				standard	z-line	0.961	wrong
				standard	**(y,z)-plane**	0.052	correct
				(y,z)-semi	point	0.961	wrong
				(y,z)-semi	z-line	0.052	correct
				z-semi	*point*	0.009	correct

The results given in Table 1 fully confirm the heuristical considerations above.

We want to point out that the convergence factors above do **not** allow one to judge which of the "correct" methods is the most efficient one for the respective situation. In order to evaluate the correct methods among each other it is, of course, necessary to take the computational effort into account!

There is, however, an essential lack of information about the computational work as long as no decision has been made how, e.g., the plane relaxation is to be performed practically. This question will be treated in the next section.

Another problem which makes any comparison of efficiency difficult are the different coarsening strategies under consideration. Clearly, standard coarsening gives an (asymptotic) ratio of $\frac{1}{8}$ for the number of coarse-grid points compared to the fine ones, but 2D- and 1D-semi-coarsening lead to ratios of $\frac{1}{4}$ and $\frac{1}{2}$, respectively. V- and W-cycles thus have essentially different complexities in these cases; only standard coarsening yields a simple recursive strategy for the definition of a sequence of coarser grids.

As we are interested only in principal results in this paper, we will therefore treat only standard coarsening in the next section.

The **proof** of "Result 1" is only sketched here. The theoretical basis is the model problem analysis as it was provided in [4] for the 2D-case. In the 3D-case, we consider the (at most) eight-dimensional spaces of gridfunctions

$$E^h_{k,l,m} = span[\sin(\pi\kappa x)\sin(\pi\lambda y)\sin(\pi\mu z) : \kappa = k, n-k;\ \lambda = l, n-l;\ \mu = m, n-m] \qquad (k,l,m = 1,2,\ldots,\frac{n}{2})$$

where the gridpoints (x,y,z) vary in Ω_h. (If one, two or three of the indices equal $\frac{n}{2}$, the dimension of $E^h_{k,l,m}$ is four, two or one, resp.) These spaces $E^k_{k,l,m}$ turn out to be invariant under all relaxation operators and all coarse-grid correction operators considered in Table 1. (The eight-dimensional spaces occur, in particular, in connection with the coarse-grid correction operator that corresponds to standard coarsening. For all other relaxation and coarse-grid correction operators, we have a still finer sub-structure of invariant spaces, namely four-dimensional subspaces for the 2D-semi-coarsenings and the alternating line relaxation and two-dimensional subspaces for the 1D-semi-coarsenings and for all the other relaxation operators considered.)

In any case we have

$$M_h^H : E^h_{k,l,m} \longrightarrow E^h_{k,l,m} \quad (k,l,m = 1,2,\ldots,\frac{n}{2}).$$

Thus, with respect to $E^h_{k,l,m}$, the restriction of M_h^H to $E^h_{k,l,m}$ has an (at most) (8,8)-matrix representation $\hat{M}^H_{h;k,l,m}$, so that

$$\rho(\hat{M}^H_{h;k,l,m})$$

can be determind by solving an (at most) (8,8)-matrix eigenvalue problem. As the collection of all $E^h_{k,l,m}$ $(k,l,m = 1,2,\ldots,\frac{n}{2})$ gives the full space of gridfunctions, we have

$$\rho(M_h^H) = max\{\rho(\hat{M}^H_{h;k,l,m}) : k,l,m = 1,2,\ldots,\frac{n}{2}\}.$$

Finally, ρ^* is obtained by

$$\rho^* = sup\{\rho(M_h^H) : h = \frac{1}{n},\ (n = 2^p,\ p = 2,3,\ldots)\}.$$

We would like to make two additional remarks.

Remark 1: Table 1 shows that there are always two strategies for obtaining good smoothings for a concrete anisotropic operator: One can invest work in the relaxation:

– cheapest: point, most expensive: plane –

or one can invest work in the coarsening strategy:

– cheapest: standard, most expensive: 1D-semi –

In Table 1 we have marked by *italics* the respective strategies with point relaxation and the "cheapest admissible" coarsening and by **boldface** the strategies with standard coarsening and the "cheapest admissible" relaxation.

Without giving corresponding results, we would like to mention, that one may, of course, always invest more work than the minimal admissible one. For example, in the case (4) one obtains also correct methods, if (y,z)-plane relaxation is combined with 2D- or 1D-coarsening, or – the other way around – if z-semi-coarsening is combined with line or plane relaxation.

Remark 2: All relaxation methods considered above (point, line and plane) are essentially parallelizable: Because of the 3D-red-black order of points, the 2D-red-black order of lines and the zebra order of planes, one relaxation step consists in all cases of *two parts* the operations of each of which can be performed in parallel.

3. USING 2D-MULTIGRID FOR PLANE RELAXATION

In this section, we want to discuss shortly the question how plane relaxation can be performed practically.

In the above analysis, it was assumed that, for each plane in each relaxation step, the corresponding 2D-problems are solved exactly. The question that arises here is whether the smoothing properties of plane relaxation deteriorate (essentially?) if the exact 2D-solution processes are replaced by approximate ones. Can, in particular, 2D-MG methods be used for this purpose? More precisely, we want to answer the following questions:

The theoretical question: A typical feature of all reasonable MG methods on non-adaptive grids is their O(N) complexity (N=total number of gridpoints). For an iterative MG method this means, that the number of operations to achieve a certain (h-independent) accuracy is proportional to the number of grid points. With respect to plane relaxation and its approximate treatment by use of 2D-MG, this property is valid if a *fixed* – h-independent – number of 2D-MG cycles (per plane) suffices in order to obtain an h-independent overall convergence factor for the resulting 3D-MG method. Can this be guaranteed?

The practical question: How should the "2D-MG approximate plane relaxation" be arranged practically (how many cycles, which kind of cycles, how many 2D-relaxations per cycle, etc.)?

In order to answer the above questions, we use a slight extension of the two-grid model problem analysis applied in Section 2. (As we consider, apart from the 3D-operator, auxiliarily also 2D-operators – for plane relaxation – in the following, we will distinguish them by corresponding prefixes.) The eight-dimensional spaces $E^h_{k,l,m}$ remain invariant under the corresponding 3D-two-grid iteration operator if in this process the "exact plane relaxation" is replaced by suitable 2D-two-grid iterations (per plane). More formally: One has

$$\widetilde{M}_h^{2h} \; : \; E^h_{k,l,m} \longrightarrow E^h_{k,l,m} \quad (k,l,m = 1,2,\ldots,\frac{n}{2})$$

where

$$\widetilde{M}_h^{2h} = \widetilde{S}_h^{\nu_2}[I_h - I^h_{2h}L^{-1}_{2h}I^{2h}_h L_h]\widetilde{S}_h^{\nu_1}$$

and $\widetilde{S}_h$ stands for one approximate plane relaxation step which consists – in its turn – of suitable 2D-two-grid iterations per plane.

The following results are obtained by this extended 3D-two-grid analysis. We want to give here only very few but characteristic results. Many more systematical investigations all of which confirm the principal behavior demonstrated here have been performed by the first author. A collection of such results will be contained in [6].

According to Result 1, there are two cases in which plane relaxation is necessary in connection with standard coarsening, namely

(3) a=1, b=100, c=100 and (4) a=1, b=100, c=10000.

The "Results" 2 and 3 refer to these cases.

Result 2: *For the case (3), Table 2.1 contains the 3D-two-grid convergence factors (h-independent upper limits)*

$$\rho^* = sup\{\rho(\widetilde{M}_h^{2h}) : h = \frac{1}{n}, \; (n = 2^p, \; p = 2,3,\ldots)\}.$$

As in Table 1, we always assume that $\nu = \nu_{3D} = (1,1)$ *plane relaxation steps are carried out.*

TABLE 2.1: 3D-two-grid convergence factors ρ^* for case (3), using only **one** 2D-two-grid step for plane relaxation (per plane)

	plane relaxation			
	exact	2D-two-grid approximate		
		$\nu_{2D} = (1,0)$	$\nu_{2D} = (1,1)$	$\nu_{2D} = (2,1)$
ρ^*	0.052	0.063	0.053	0.053

Here, the figures have the following meaning:

exact: The plane relaxation is carried out exactly (cf. Table 1).

2D-two-grid approximate: The plane relaxation is carried out approximately by performing **one** step of a 2D-two-grid method for each plane. In this 2D-two-grid method we use 2D-red-black pointwise relaxation for smoothing, namely

$$\nu_{2D} = (1,0), \quad (1,1), \text{ or } (2,1).$$

The 2D-coarse-grid operator is the one canonically obtained from the 2D-fine grid operator. For the 2D-grid transfers full weighting and bilinear interpolation are used.

From this table we see that *one 2D-two-grid step with* $\nu_{2D} = (1,1)$ *relaxations per plane* is sufficient to obtain the same overall 3D-two-grid convergence factor as that of an exact "outer" plane relaxation step. Clearly, additional inner relaxation steps ($\nu_{2D} = (2,1)$) can give no better overall asymptotic convergence.

In Table 2.2, we give some interesting results from *numerical experiments*. These refer to the case $h = \frac{1}{64}$. We give numerically measured asymptotic convergence factors for different types of cycles: F- and V-cycles for the outer 3D-MG iteration are combined with F- and V-Cycles for the **one** inner 2D-MG iteration that is used for plane relaxation (per plane).

TABLE 2.2: 3D convergence factors ρ_{exp} measured experimentally, for $h = \frac{1}{64}$

$\rho_{exp},\ h = \frac{1}{64}$	2D-MG approximate plane relaxation			
	$\nu_{2D} = (1,0)$		$\nu_{2D} = (1,1)$	
	2D-MG-F	2D-MG-V	2D-MG-F	2D-MG-V
3D-MG-F	0.063	0.097	0.049	0.053
3D-MG-V	0.070	0.097	0.070	0.070
ρ^*	0.063		0.053	

We do not intend to interpret these results in all details. We only want to point out, that the use of (2D- and 3D-) F-cycles always yields the best achievable result ($\rho_{exp} \leq \rho^*$), as far as convergence factors are concerned. However, with respect to efficiency, still simpler algorithms (V-cycles) may be advantageous.

Similar results as listed above for case (3) are valid for case (4). In analogy to Result 2, we will formulate a corresponding Result 3 below. We want to point out, however, one difference here explicitly.

The 2D-problem which has to be solved in each (y,z)-plane relaxation step in case (4) is an anisotropic one (b=100, c=10000). Therefore, line relaxation has to be employed for smoothing in the 2D-MG process! This means, that we have – in some sense – a threefold MG-recursion in this case: Smoothing in the 3D-MG iteration consists of a 2D-MG process the smoothing of which consists of a "1D-MG process " (1D-cyclic reduction or tridiagonal LU-decomposition). At first sight, this may appear complicated. In fact, it can be implemented very simply and performed very cheaply.

Result 3: *For the case (4), Table 3.1 contains 3D-two-grid convergence factors, namely the h-independent limits*

$$\rho^* = sup\{\rho(\widetilde{M}_h^{2h}) : h = \frac{1}{n},\ (n = 2^p,\ p = 2,3,\ldots)\}$$

and

$$\rho(\widetilde{M}_h^{2h}) \text{ for } h = \frac{1}{64}.$$

TABLE 3.1: 3D-two-grid convergence factors for case (4), using **one** 2D-two-grid step for plane relaxation (per plane)

	plane relaxation		
	exact	2D-two-grid approximate	
		$\nu_{2D} = (1,0)$	$\nu_{2D} = (1,1)$
ρ^*	0.052	0.053	0.053
$\rho(\widetilde{M}_h^{2h}), h = \frac{1}{64}$	0.007	0.015	0.007

Here, the figures have an analogous meaning as explained for Table 2.1.

Again, we observe that one *2D-two-grid step with $\nu_{2D} = (1,1)$ relaxation steps per plane* is sufficient to obtain the same overall 3D-two-grid convergence factor as that of an exact "outer" plane relaxation step. Practically, already $\nu_{2D} = (1,0)$ relaxation step per plane will do in this case.

Furthermore, from Table 3.1 we recognize an essential difference between the h-independent upper limits ρ^* and the corresponding convergence factors for fixed $h = \frac{1}{64}$. (Note, that both values are based on the model problem analysis sketched above!) An h-dependency as observed here occurs sometimes in highly anisotropic problems which is well-known already from the 2D-case. One would not find it in isotropic model problems.

In analogy to Table 2.2, we give some results from *numerical experiments* in Table 3.2.

TABLE 3.2: 3D convergence factors ρ_{exp} measured experimentally, for $h = \frac{1}{64}$

ρ_{exp}, $h = \frac{1}{64}$	2D-MG approximate plane relaxation $\nu_{2D} = (1,0)$	
	2D-MG-F	2D-MG-V
3D-MG-F	0.015	0.015
3D-MG-V	0.015	0.015
$\rho(\widetilde{M}_h^{2h})$, $h = \frac{1}{64}$	0.015	

Obviously this table is somewhat boring: Already the simplest and cheapest approach (only one 2D-MG-relaxation step ($\nu_{2D} = (1,0)$), V-cycles for the inner 2D-MG and the outer 3D-MG) yields the best achievable result ($\rho_{exp} \leq \rho(\widetilde{M}_h^{2h})$).

Finally, we want to make some remarks about computational work and efficiency. Systematical studies of these questions can be found in [6].

For each of the cases (1)-(4), we consider in Table 4 simple MG versions of those "correct" two-grid methods in *Table 1* which are based on *standard coarsening*.

TABLE 4: Computational work versus convergence factors

case	PWU	ρ_{exp} $(h = \frac{1}{64})$	case	LWU	ρ_{exp} $(h = \frac{1}{64})$
(1)	$\sim 3\frac{1}{2}$	0.23	(2)	$\sim 3\frac{1}{2}$	0.080
(3)	$\sim 7\frac{1}{2}$	0.097	(4)	$\sim 7\frac{1}{2}$	0.015

These figures refer to 3D-MG methods with the following characteristic components:

(1) 3D-red-black point relaxation, $\nu_{3D} = (1,1)$, V-cycle for 3D-MG
(2) 3D-line relaxation in 2D-red-black order, $\nu_{3D} = (1,1)$, V-cycle for 3D-MG
(3) plane relaxation (zebra order), $\nu_{3D} = (1,1)$, V-cycle for 3D-MG; one 2D-MG step per plane, 2D-red-black relaxation, $\nu_{2D} = (1,0)$, V-cycle for 2D-MG
(4) plane relaxation (zebra order), $\nu_{3D} = (1,1)$, V-cycle for 3D-MG; one 2D-MG step per plane, 2D-zebra-line relaxation, $\nu_{2D} = (1,0)$, V-cycle for 2D-MG.

The computational work is measured very roughly by PWU and LWU (point and line work units, resp.). In the cases (1) and (3) one PWU is defined as the work needed for one step of 3D-red-black point relaxation. In the cases (2) and (4) one LWU is defined as the work needed for one step of 3D-line relaxation (2D-red-black order).

From Table 4 we see that the methods (3) and (4) where plane relaxation is employed are only twice as expensive as comparable methods with point (1) or line (2) relaxation, respectively. The convergence properties are even better (essentially better in the case (3) compared to (1)). (The convergence factors ρ_{exp} were experimentally measured for $h = \frac{1}{64}$.)

CONCLUSIONS

In accordance with the growing significance of 3D-models in many fields of scientific computation, many users of 2D-multigrid have started to deal with 3D-multigrid. However, although plane relaxation has been recognized to be necessary for smoothing in many practically important 3D-cases, very little substantial work has been done so far in this direction. Even some of those who have dealt with plane relaxation by now obviously do not believe in its practical feasibility (see e.g. [2]).

This paper shows that plane relaxation is a necessary, powerful and cheap tool for smoothing in 3D-multigrid.

REFERENCES

[1] Brandt, A.: *Multigrid Techniques: 1984 Guide with Applications to Fluid Dynamics.* GMD-Studie Nr. 85, GMD, St. Augustin, 1984.

[2] Behie, A.; Forsyth Jr., P. A.: *Multi-grid solution of three dimensional problems with discontinous coefficients.* Appl. Math. and Comp., **13**, pp. 229–240, 1983

[3] Hackbusch, W.; Trottenberg, U. (eds.): *Multigrid Methods. Proceeding of the Conference Held at Köln-Porz, November 23-27, 1981.* Lecture Notes in Mathematics, 960. Springer-Verlag, Berlin, 1982.

[4] Stüben, K.; Trottenberg, U.: *Multigrid methods: fundamental algorithms, model problem analysis and applications.* In [3], pp. 1–176.

[5] Stüben, K.; Trottenberg, U.; Witsch, K.: *Software development based on multigrid techniques.* In PDE Software, Modules, Interfaces and Systems (B. Engquist, T. Smedsaas, eds.).North-Holland, Amsterdam, 1984.

[6] Thole, C.A.: *Efficient multigrid methods for standard elliptic 3D-problems, local Fourier analysis and numerical experiments.* To appear.

A PRECONDITIONED CONJUGATE RESIDUAL ALGORITHM FOR THE STOKES PROBLEM

R. Verfürth *

INRIA Domaine de Voluceau - Rocquencourt

B.P. 105 78153 LE CHESNAY Cedex

I. INTRODUCTION

We present a preconditioned conjugate residual algorithm for a mixed finite element discretization of the Stokes problem

$$\begin{aligned} -\Delta \underline{u} + \nabla p = \underline{f} \quad \text{in } \Omega \quad &, \quad \underline{u} = 0 \quad \text{on } \partial\Omega \\ \operatorname{div} \underline{u} = 0 \quad \text{in } \Omega & \end{aligned} \tag{1.1}$$

in a plane polygonal domain Ω . The preconditioning relies on the idea of hierarchical basis functions for finite elements (cf.[7,8]). The algorithm has a quasi optimal convergence rate of $1-O(|\log h|)$.

Numerical experiments yield convergence rates of .9 - .96. Despite these rather poor convergence rates, the algorithm is competitive with other iterative methods for the Stokes problem (Cf. [4, 5, 6])because of its low cost. This is supported by two examples of Navier-Stokes calculations. For a more detailed comparison of different iterative methods for the Stokes problem including multigrid methods we refer to [6].

2. PRELIMINARIES

Let J_{h_j} , $j = 0, 1, \ldots, R$, be a sequence of regular triangulations of Ω with $h_j = \frac{1}{2} h_{j-1}$, $j = 1, \ldots, R$ (cf. [1]). We assume that each $\mathcal{T} \in J_{h/2}$ is obtained by dividing an appropriate $\mathcal{T}^1 \in J_h$ into four equal triangles the vertices of which are the midpoints of sides and the vertices of $\mathcal{T}^1$. Denoting by S_j the space of continuous piecewise linear finite elements corresponding to J_h , we put

$$S_j^o := \{ \varphi \in S_j : \varphi = 0 \text{ on } \partial\Omega \}, \; X_j := S_j^o \times S_j^o ,$$

$$M_j := \{ p \in S_{j-1} : \int_\Omega p\,dx = 0 \}, \; 1 \leq j \leq R,$$

Note that M_j corresponds to a coarser triangulation than X_j.

Permanent address: Mathematisches Institut der RUB, Universitätsstr. 150
4630 Bochum, W. - GERMANY

Denote by

$$\|\varphi\|_k := \{\int_\Omega \sum_{|\alpha|\leq k} |D^\alpha \varphi|^2 \, dx \}^{1/2}, \quad k \geq 0,$$

the usual norm of the Sobolev space $H^k(\Omega)$ and by $(.,.)$ the scalar product of $L^2(\Omega)$. We use the same notation for the corresponding scalar products and norms on product spaces. Finally, we introduce the continuous bilinear forms

$$a(u,v) := \int_\Omega \nabla \underline{u} \, \nabla \underline{v} \, dx, \quad b(\underline{v},p) := -\int_\Omega p \, \mathrm{div} \, \underline{v} \, dx, \quad (2.2)$$

$\mathscr{L}([\underline{u},p],[\underline{v},q]) := a(u,\underline{v}) + b(\underline{v},p) + b(\underline{u},q)$ on $X_j \times X_j$, $X_j \times M_j$ and $(X_j \times M_j) \times (X_j \times M_j)$ resp.

Then the mixed finite element approximation of the Stokes problem (1.1) is to find $[\underline{u}_j, p_j] \in X_j \times M_j$ such that

$$\mathscr{L}([u_j, p_j], [\underline{v}_j, q_j]) = (\underline{f},\underline{v}_j) \; \forall [\underline{v}_j, q_j] \; X_j \times M_j, \; 1 \leq j \leq R. \; (2.3)$$

Actually, we want to solve (2.3) only on the finest level $j=R$. The coarser triangulations are only auxiliary ones.

A crucial point for our analysis is the fact that the described discretization of (1.1) satisfies the so called Babuška - Brezzi condition:

$$\inf_{[\underline{u},p] \in X_j \times M_j} \; \sup_{[\underline{v},q] \in X_j \times M_j} \frac{\mathscr{L}([\underline{u},p],[\underline{v},q])}{\{\|u\|_1^2 + \|p\|_0^2\}^{1/2} \{\|\underline{v}\|_1^2 + \|q\|_0^2\}^{1/2}}$$

$$\geq \gamma > 0 \quad (2.4)$$

with a constant γ independent of h. This is a consequence of the stability result of [3] . Inequ. (2.4) ensures the unique solvability of (2.3) and implies optimal error estimates for the finite element approximation [2,3].

3. PRECONDITIONING OF THE DISCRETE STOKES PROBLEM

Denote by Ω_j the set of vertices of triangles in J_{h_j} and by $I_j : C(\bar{\Omega}) \to S_j$ the usual interpolation operator. We define a mesh dependent scalar product on S_R^o by

$$((\varphi,\psi))_{1,R} := (\nabla I_o \varphi, \nabla I_o \psi) + \sum_{j=1}^{R} \sum_{x \in \Omega_j | \Omega_{j-1}} [I_j \varphi - I_{j-1}\varphi] \cdot [I_j \psi - I_{j-1}\psi](x) \quad (3.1)$$

The corresponding norm is denoted by $|||\cdot|||_{1,R}$. We use the same notation for the corresponding scalar product and norm on the product space X_R. The following result is proved in [7,8] :

Proposition 3.1 : There are two constants $0 < C_0^* < C_1^*$ which do not depend on h_o, h_R such that

$$C_0^*(R+1)^{-2} \|\varphi\|_1 \leq |||\varphi|||_{1,R} \leq C_1^* \|\varphi\|_1 \quad \forall \varphi \in S_R^o. \qquad (3.2)$$ □

Equations (2.4) and (3.2) immediately imply :

Corollary 3.2 : There are two constants $0 < C_0 < C_1$ which do not depend on h_o, h_R such that

$$\inf_{[\underline{u},p] \in X_R \times M_R} \sup_{[\underline{v},q] \in X_R \times M_R} \frac{\mathscr{L}([\underline{u},p], [\underline{v},q])}{\{|||\underline{u}|||^2_{1,R} + \|p\|_0^2\}^{1/2} \{|||\underline{v}|||^2_{1,R} + \|q\|_0^2\}^{1/2}}$$

$$\geq C_o(R+1)^{-2} \quad (3.3)$$

and

$$\mathscr{L}([\underline{u},p], [\underline{v},q]) \leq C_1 \{|||\underline{u}|||^2_{1,R} + \|p\|_0^2\}^{1/2} \{|||\underline{v}|||^2_{1,R} + \|q\|_0^2\}^{1/2} \quad (3.4)$$

$$\forall [\underline{u},p], [\underline{v},q] \in X_R \times M_R.$$ □

The crucial point is to interprete Corollary 3.2 as a preconditioning result for the discrete Stokes problem (2.3). To this end, we have to introduce the notation of hierarchical basis functions for S_R^o [7,8]. Denote by ψ_j^i, $0 \leq j \leq R$, $1 \leq i \leq N_j := \dim S_j^o$ the hierarchical basis. It is defined recursively : If $j = 0$, the ψ_j^i are the usual nodal basis functions of S_o^o ; If $j \geq 1$, the ψ_j^i consist of the ψ_{j-1}^i plus the nodal basis functions corresponding to $\Omega_j \setminus \Omega_{j-1}$. Figure 1 shows the nodal and hierarchical basis for a partition of [0,1] corresponding to $j = 2$. In the sequel, we use the convention that for any basis of $X_R \times M_R$ we number first the basis functions for the x component of the velocity, then those for its y component and finally those for the pressure .

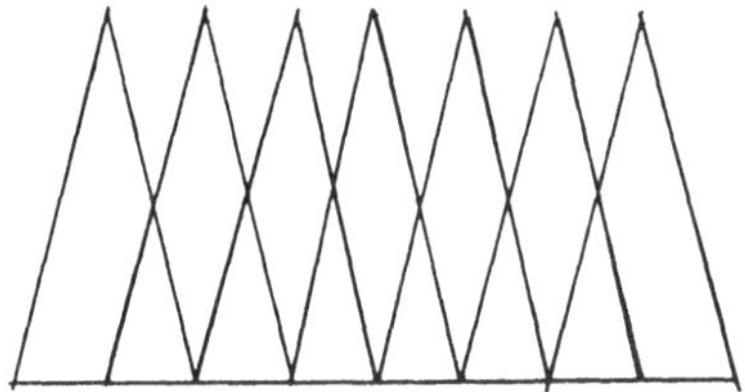

Fig. 1 : nodal basis

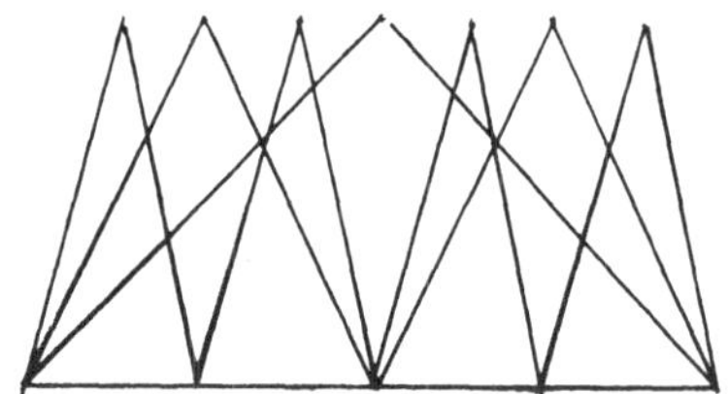

hierarchical basis

Denote by

$$A = \begin{pmatrix} A_{11} & 0 & A_{13} \\ 0 & A_{22} & A_{23} \\ A_{13}^T & A_{22}^T & 0 \end{pmatrix}$$

the stiffness matrix of $\mathcal{L}$ with respect to the nodal basis. Let $\tilde{S}$ be the transformation matrix from the hierarchical to the nodal basis of S_R^o and put

$$S := \begin{pmatrix} \tilde{S} & 0 & 0 \\ 0 & \tilde{S} & 0 \\ 0 & 0 & I \end{pmatrix} .$$

Finally, denote by B the stiffness matrix of the scalar product $((\underline{u},\underline{v}))_{1,R} + (p,q)$ on $X_R \times M_R$ with respect to the hierarchical/nodal basis of X_R and M_R resp. Note that

$$B = \begin{pmatrix} \tilde{B} & 0 & 0 \\ 0 & \tilde{B} & 0 \\ 0 & 0 & I \end{pmatrix} ,$$

where $\tilde{B}$ is a diagonal matrix except a small diagonal block corresponding to $(\underline{\nabla}\, I_o\, \underline{u}, \nabla\, I_o\, \underline{v})$. Since B is symmetric and positive definite, it can be decomposed as $B = LL^T$ with a lower triangular matrix L. Put

$$Q := S^{-T} L. \tag{3.5}$$

Then Corollary 3.2 implies

$$C_o(R+1)^{-2} \leq \inf_{X \in R^k} \sup_{Y \in R^k} \frac{X^T S^T A S Y}{\{X^T LL^T X\}^{1/2} \{Y^T LL^T Y\}^{1/2}}$$

$$\leq \sup_{X \in R^k} \sup_{Y \in \mathbb{R}^k} \frac{X^T S^T A S Y}{\{X^T LL^T X\}^{1/2} \{Y^T LL^T Y\}^{1/2}}$$

$$\leq C_1.$$

This proves :

<u>Proposition 3.3</u> : The condition number of $Q^{-1} A Q^{-T}$ is bounded by $\frac{C_1}{C_0} (R+1)^2 \sim |\log h|^2$. A conjugate residual algorithm applied to the discrete Stokes problem (2.3) with QQ^T as preconditioning matrix has the quasi optimal convergence rate $1-0(|\log h|)$. □

In each step of the preconditioned conjugate residualalgorithm, we have to calculate the solution of

$$QQ^T\, Y = Z \tag{3.6}$$

with a suitable $Z \in \mathbb{R}^k$. This can be done by the following three steps :

(i) transform the velocity components of Z into their hierarchical basis repesentation. Denote the resulting vector by $Y^{(1)}$ and by w its velocity components corresponding to the coarsest space X_o.

(ii) Solve the discrete analogue of $\Delta v = w$ on X_o. Replace w in $Y^{(1)}$ by v and denote the resulting vector by $Y^{(2)}$.

(iii) Transform the velocity components of $Y^{(2)}$ into their nodal basis representation. The result is the solution y of (3.6).

Note that the pressure components remain unchanged in the above calculation. The cost for solving (3.6) corresponds to the calculation of about three scalar products. Hence the preconditioning is very cheap.

4. NUMERICAL RESULTS

We consider the following three regions :

(i) the unit square $\Omega_c := (0,1) \times (0,1)$,

(ii) the L-shaped domain $\Omega_L := \Omega_c \setminus [0.5,1] \times [0.5,1]$,

(iii) the slit unit square $\Omega_S := \Omega_c \setminus [0.5,1] \times [0.5]$.

The triangulation is made up by isosceles, right angled triangles with short sides of length h. We consider the meshsizes h = 1/4, 1/8, 1/16, 1/32. The coarsest mesh is h = 1/4 for Ω_c and h = 1/8 for Ω_L and Ω_S. We have tested the algorithm for five examples. In the first two examples, the exact solution is known :

Ex 1 : $\underline{u}(x,y) := (200\, x^2(1-x)^2 y\,(1-y)\,(1-2y),$
$\qquad -200\, x\,(1-x)\,(1-2x)\, y^2\,(1-y)^2)$
$p(x,y) := 100\, x\,(1-x)\, y\,(1-y) - \frac{25}{9}$

Ex 2 : $\underline{u}(x,y) := (2\,\pi\,\mathrm{Sin}^2\,(2\,\pi x)\,\mathrm{Sin}\,(4\,\pi\, y),$
$\qquad -2\,\pi\,\mathrm{Sin}\,(4\,\pi x)\,\mathrm{Sin}^2(2\,\pi\, y))$
$p(x,y) := 4\pi^2\,\mathrm{Sin}\,(4\pi\, x)\,\mathrm{Sin}\,(4\,\pi\, y).$

In the last three examples, the exact solution is unknown. The right hand side is given by :

Ex. 3 : $\underline{f}(x,y) := \underline{e} := (1,-1)$,
Ex. 4 : $\underline{f}(x,y) := 100\, x\,(1-x)\, y\,(1-y)\,\underline{e}$,
Ex. 5 : $\underline{f}(x,y) := 100 \exp\,[-\,100\,(x^2+y^2)]\,\underline{e}$.

Because of the boundary conditions, example 1 is considered only on Ω_C. The conjugate residual algorithm is restarted every 10 iterations. For examples 1, 2, the iteration terminates if the difference between the calculated and the exact pressure is less than 0.1 h in the L^2 -norm. The factor h reflects the error estimate of O(h) for the finite element approximation [3]. For the other examples, the iteration is terminated if the L^2 - norm of the residual is less than 10^{-4}. The starting value on the coarsest grid is zero. On the finer grids, it is the interpolate of the last iterate on the next coarser grid. The calculations were done on the Bull C II - HB - DPS 68/Multics at INRIA.

Tables 1, 2 give the mean convergence rates of the algorithm for the five examples.

Table 1 : Mean convergence rates for known exact solution

h^{-1}	Ex. 1 Ω_C	Ex. 2 Ω_C	Ex. 2 Ω_L	Ex. 3 Ω_S
8	.94	-	-	-
16	.83	.94	.94	.93
32	.87	.95	.94	.94

Table 2 : mean convergence rates for given $\underline{f}$

h^{-1}	Ex. 3 Ω_C	Ex. 3 Ω_L	Ex. 3 Ω_S	Ex. 4 Ω_C	Ex. 4 Ω_L	Ex. 4 Ω_S	Ex. 5 Ω_C	Ex. 5 Ω_L	Ex. 6 Ω_S
8	.97	.97	.96	.97	.97	.97	.96	.97	.97
16	.95	.94	.94	.94	.97	.95	.96	.97	.97
32	.94	.93	.95	.95	.95	.95	.97	.96	.96

The results show that the geometry of Ω has no influence on the convergence rates. This is in accordance with the convergence analysis which (in contrast to multigrid methods [5]) does not require any regularity assumptions.
At first sight, convergence rates of .9-.96 seem rather poor. For the same examples, we obtained in [4] convergence rates of .8 - .9 using an Uzawa-type algorithm which combines a conjugate gradient and a multigrid method. First numerical experiments with the multigrid algorithm of [5] yield convergence rates of about .4 - .5. However, the costs of these algorithms are quite different. An operation count shows that the preconditioned conjugate residual algorithm is five times cheaper then the multigrid algorithm. This is supported by the numerical experiments where the preconditioned conjugate residual algorithm needed 25 % CPU time less than the Uzawa type algorithm. The difference is even more striking for Navier-Stokes calculations. We used both algorithms as Stokes solvers in a fixed point iteration for the driven cavity problem and the flow across a step at Reynolds numbers between 1o and 50. The preconditioned conjugate residual algorithm needed four times as many linear iterations than the Uzawa-type algorithm. The computing time, however, was only one half. Depending on the Reynolds number, between 1.5 and 3 minutes were needed to calculate the solution for $h = 1/32$ starting from $h = 1/8$. These examples show that the presented algorithm is an alternative to other iterative Stokes solvers. For a more detailed comparison of the different methods, we refer to [6].

References

[1] Ph. G. CIARLET : "The finite element method for elliptic problems". North Holland, Amsterdam, 1978.

[2] V. GIRAULT, P.A. RAVIART : "The finite element approximation of the Navier-Stokes equations". Springer, Berlin. 1979.

[3] R. VERFÜRTH : "Error estimates for a mixed finite element approximation of the Stokes equations". RAIRO 18, 175-182 (1984).

[4] R. VERFÜRTH : " A combined conjugate gradient-multigrid algorithm for the numerical solution of the Stokes problem". IMA J. Numer. Anal. 4, 441-455 (1984).

[5] R. VERFÜRTH : "A Multi-Level algorithm for mixed problems". SIAM J. Numer. Anal. 21, 264-271 (1984).

[6] R. VERFÜRTH : "Iterative methods for the numerical solution of mixed finite element approximations of the Stokes problem". Report INRIA, 1985.

[7] H. YSERENTANT : "On the multi-Level splitting of finite element spaces". Bericht Nr. 21 RWTH Aachen, 1983.

[8] H. YSERENTANT : "Über die Aufspaltung von finite Element Räumen in Teilräume verschiedener Verfeinerungsstufen". Habilitationsschrift, RWTH Aachen, 1984.

Participants

Alt, H.W., *Institut für Angewandte Mathematik, Universität Bonn, Wegelerstr. 6, D-5300 Bonn 1.*

Axelsson, O., *Dept. of Mathematics, University of Nijmegen, Toernooiveld, NL-6525 ED Nijmegen.*

Bank, R.E., *Dept. of Mathematics, University of California at San Diego, La Jolla, CA 92093, USA.*

Böhmer, K., *Fachbereich Mathematik, Universität Marburg, Lahnberge, D-3550 Marburg.*

Bollrath, Chr., *Mathematisches Institut, Ruhr-Universität Bochum, Universitätsstr. 150, D-4630 Bochum 1.*

Braess, D., *Abt. f. Mathematik d. Universität, Universitätsstr. 150, D-4630 Bochum-Querenburg.*

Brandt, A., *Dept. of Mathematics, Weizmann Inst. of Science, P.O. Box 26, Rehovot, Israel.*

Bulirsch, R., *TU München, Fakultät f. Math. u. Informatik, Arcisstr. 21, D-8000 München 2.*

Hackbusch, W., *Inst. f. Informatik und Prakt. Mathematik der Universität Kiel, Olshausenstr. 40, D-2300 Kiel 1.*

Hemker, P.W., *Mathematisch Centrum, P.O. Box 4079, NL-1009 AB Amsterdam.*

Hofmann, G., *Institut für Informatik, Christian-Albrechts-Universität Kiel, Olshausenstr. 40, D-2300 Kiel 1.*

Holstein, H., *Dept. of Computer Science, University College of Wales, Aberystwyth SY23 3BZ, Great Britain.*

Karlisch, J., *Inst. f. Mathematik der Ruhr-Universität, Universitätsstr. 150, D-4630 Bochum 1.*

Kettler, R., *Van Hasseltlaan 194, NL-2625 HL Delft.*

Lacor, C., *Vrije Universiteit Brussel, Dept. of Fluid Mechanics, Pleinlaan 2, B-1050 Brussels.*

Lazarov, R.D., *Inst. of Mathematics, Bulgarian Academy of Science, Sofia 1113, Bulgarien.*

Linden, J., *GMD-IMA, Postfach 1240, D-5205 St. Augustin 1.*

Maitre, J.F., *Mathematiques, Ecole Centrale de Lyon, F-69130 Ecully.*

Mandel, J., *Matematieko-Fysikálni Fakulta, University Karlovy, Malostranské nam. 25, CS-11800 Prag 1.*

Mantel, B ., *D E A , Avions Maral Dassault / BA, 78 Quai Carnot, F-92210 St. Cloud.*

Marek, I., *Matematieko-Fysikálni Fakulta, University Karlovy, Malostranské nam. 25, CS-11800 Prag 1.*

McCormick, S.F., *Dept. of Mathematics, University of Colorado of Denver, 14th Street 1100, Denver CO 80208, USA.*

Merriam, M.L., *Mail Stop 202 A-1, NASA Ames Research Center, Moffett Field, CA 94035, USA.*

Mittelmann, H.D., *Dept. Math., Arizona State University, Tempe, AZ 85282, USA.*

Musy, F., *MIS, Ecole Centrale de Lyon, BP 163, F-69131 Ecully Cedex.*

Nitsche, J., *Institut für angewandte Mathematik, Hermann-Herder-Str. 10, D-7800 Freiburg.*

Peisker, P., *Institut für Mathematik, Ruhr-Universität Bochum, Universitätsstr. 150, D-4630 Bochum 1.*

Periaux, J., *D E A , Avions Maral Dassault/BA, 78 Quai Carnot, F-92210 St. Cloud.*

Pitkäranta, J., *Institute of Mathematics, Helsinki University of Technology, SF-02150 Espoo 15.*

Qun, L., *Academia Sinica, Dept. of System Sciences, Beijing 100080 China.*

Rannacher, R., *Fachbereich Angewandte Mathematik und Informatik (10), Universität des Saarlandes, D-6600 Saarbrücken.*

Ruge, J., *Dept. of Mathematics, University of Colorado of Denver, 14th Street 1100, Denver CO 80208, USA.*

Shao, X., *Inst. f. angewandte Mathematik, Gebäude 293, D-6900 Heidelberg.*

Steffen, B., *KFA/ZAM, Postfach 1913, D-5170 Jülich.*

Stetter, H.J., *Inst. f. Angew. u. Num. Math., TU Wien, Gusshausstr. 27-29, A-1040 Wien.*

Stüben, K., *GMD-F1/T, Postfach 1240, D-5205 St. Augustin 1.*

Trottenberg, U., *GMD-F1/T, Postfach 1240, D-5205 St. Augustin 1.*

Verfürth, R., *Institut für Mathematik, Ruhr-Universität Bochum, Universitätsstr. 150, D-4630 Bochum 1.*

Wesseling, P., *Onderafdeling der Wiskunde en Informatica, Technische Hogeschool Delft, Julianalaan 132, NL-2628 BL Delft.*

Witsch, K., *Mathematisches Institut, Universität Düsseldorf, Universitätsstr. 1, D-4000 Düsseldorf.*

Wittum, G., *Inst. f. Inform. u. Prakt. Mathematik, Christian-Albrechts-Universität Kiel, Olshausenstr. 40, D-2300 Kiel 1.*

Yserentant, H., *Inst. f. Geometrie u. Prakt. Mathematik, RWTH Aachen, Templergraben 55, D-5100 Aachen.*